Upon General Relativity

How space-time, gravity,
and electromagnetism
emerge from the
spinor algebras

Dennis Morris

(August 2015)

Published by: Abane & Right

31/32 Long Row

Port Mulgrave

Saltburn

TS13 5LF

01287 678918

August 2015

Contents

Contents

Contents

Chapter 1

An Overview of this Book

We will derive our 4-dimensional space-time and both general relativity and classical electromagnetism from the six A_3 spinor algebras. (Roughly, the A_3 spinor algebras are Clifford algebras.) To derive the classical universe, we will use three operations.

The first operation is super-imposition of the six A_3 spinor algebras to form the emergent classical universe. This is just taking the expectation space, expectation tensor, and expectation field equations by adding the six spaces, six tensors, and six sets of field equations in the six A_3 spinor algebras. Super-imposition is the generalisation of the operation of taking expectation values in quantum mechanics. It is by super-imposition that we move from the quantum universe to the classical universe. In a sense, the classical universe is no more than an average of all the A_3 spinor universes.

The second operation we call fabrication. Fabrication is building our 4-dimensional space-time from

a) a 4-dimensional locally flat expectation manifold
b) a 4-dimensional expectation distance function
c) a set of six 2-dimensional spinor spaces
d) six 2-dimensional angles within the 2-dimensional spinor spaces

and building a 4-dimensional inner product from a set of 2-dimensional inner products. We will see that this fabrication is controlled by the expectation distance function (the average of the spinor distance functions).

The third operation is to allow the phase of the A_3 spinor algebras to vary from point to point over the expectation manifold; this is a familiar operation in QFT. In technical language, we form a fibre bundle with locally varying phase.

The details of these operations will be explained in the body of this book.

Concisely, the classical physics of the universe emerges from the A_3 spinor algebras as:

1) Super-imposition of the six A_3 spinor spaces produces an emergent 4-dimensional real manifold with local flatness. Because there are six (any number more than one would be enough) A_3 algebras, the super-imposition operation destroys the algebraic structure of these algebras, and so we have no 4-dimensional spinor rotations in our classical universe.

2) Super-imposition of the six A_3 spinor distance functions produces the 4-dimensional expectation distance function of our space-time.

3) Because there is only one copy of each of the two 2-dimensional spinor algebras, the super-imposition of the two 2-dimensional spinor algebras, $\mathbb{C} \& \mathbb{S}$, leaves the 2-dimensional algebras unchanged. This means we have two types of 2-dimensional rotations in our 4-dimensional classical universe. These are Euclidean rotation and space-time rotation (Lorentz boost rotation).

4) The nature of the expectation distance function allows that the 2-dimensional rotations exist in the classical universe and that there are three of each type of the 2-dimensional rotations. The independence of the variables in the expectation distance function fabricates (fits together) these six 2-dimensional rotations into our 4-dimensional space-time.

5) The 2-dimensional spinor algebras contain 2-dimensional inner products. The independence of the variables in the expectation distance function fabricates (fits together) six 2-dimensional inner products into a 4-dimensional 4-vector inner product.

6) The 4-dimensional inner product leads to the metric tensor.

7) Because the emergent 4-dimensional space-time is not a division algebra space (spinor space), there is no algebraic structure to 'keep it flat'. Thus, we can have curvature in the

emergent 4-dimensional space-time. Note that all spinor spaces are so inexorably flat that they do not even have zero curvature.

8) An A_3 phase locally varying over the emergent manifold induces an affine connection, a sense of parallel transport, into the manifold. This is the Levi-Civita connection.

9) From the metric tensor, we get the Riemann tensor, the Ricci tensor and the Einstein tensor.

10) Super-imposition of the 1st non-commutative differentials of six A_3 potentials gives an expectation field tensor which splits into symmetric and anti-symmetric parts. The anti-symmetric part is the electromagnetic tensor. The symmetric part is the energy-momentum tensor[1].

11) Super-imposition of the 2nd non-commutative differentials of the six A_3 potentials gives the expectation field equations. The anti-symmetric part of these equations are the Maxwell equations of classical electromagnetism. The symmetric part of these equations would form gravito-electromagnetism, but they cancel in the classical universe.

12) We form GR as Einstein did from the equivalence principle and by putting the Einstein tensor equal to the energy-momentum tensor.

13) The six A_3 algebras come in three pairs. The charge of the symmetric part of the A_3 algebras is mass, and so we predict three masses for each quantum particle – three generations of matter.

We will see that, as well as our 4-dimensional, space-time emerging from the A_3 spinor algebras, Riemann geometry and Riemann mathematics also emerge from the A_3 spinor algebras. The remainder of this book is just the details of the above.

---

[1] Your author suspects that the symmetric part is the gravitational field tensor to match the electromagnetic field tensor. Conventionally, it is assumed that the energy-momentum tensor is the gravitational field tensor. Perhaps this is wrong.

Before we can produce GR, we need to familiarise the reader with Riemann geometry and with GR. The first part of the book, by far the longest part, does this.

Note: Within this book, we use the phrase 'spinor algebra'. A spinor algebra is a division algebra; these two phrases refer to the same thing. A spinor algebra (division algebra) is a type of complex numbers such as the complex numbers, $\mathbb{C}$, or the hyperbolic complex numbers, $\mathbb{S}$, or the quaternions, or the A_3 algebras or the[2] etc.. These algebras derive from the finite groups.[3]

We also use the phrase 'spinor space'. A spinor space is the same as a spinor algebra in the same way that the complex plane, $\mathbb{C}$, is the same thing as the complex numbers, $\mathbb{C}$.

---

[2] See: Dennis Morris - The Naked Spinor
[3] See: Dennis Morris - Complex Numbers The Higher Dimensional Forms

Chapter 2

An Overview of the Theory of General Relativity

The theory of general relativity, usually abbreviated to GR, was published by Albert Einstein (1879-1955) in 1915. It is a theory of the gravitational force which replaced the gravitational theory of Isaac Newton (1642-1727).

There is a fundamental difference between the way Isaac Newton viewed space and time and the way that GR views space and time. Newton thought of time as a 1-dimensional space completely separate from the 3-dimensional space we see around us. He took the view that there was a copy of our 3-dimensional space fixed to every point of the 1-dimensional space that is time. Newton viewed space and time as what mathematicians call a fibre bundle. A fibre bundle is a number (could be one) of separate types of space fixed on to an underlying space at every point of the underlying space. There is nothing that says the two, or more, spaces that are tied together in a fibre bundle have to be of any particular dimension.

Because we are artistically disadvantaged, we illustrate Newton's view with a 2-dimensional plane instead of the 3-dimensional space:

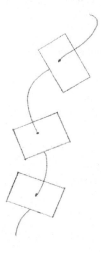

GR is based on the view that 3-dimensional space and 1-dimensional time are not two separate spaces but together form a single 4-dimensional space called space-time. This is a fundamentally different view of space and time from Newton's view of space and time as two separate spaces.

> *Aside:* It is very striking that something such as the orbits of the planets around the sun can be, to very great accuracy, described by two man-made theories based on such utterly different fundamental concepts.
>
> If we are able to produce a man-made 'theory of everything', how will we know that it is the only correct man-made theory that describes everything? Of course, if, as your author is seeking to do, we can derive everything from no more than the existence of the number one, then we can argue that the theory is not man-made and must be the true theory.
>
> *Aside:* Remarkably, the fibre bundle view underpins modern quantum field theory, QFT. In modern QFT, we let one space float over an underlying space in such a way that an angle in the floating space varies from point to point of the underlying space and this produces, using three different spaces, the photon field, the weak force, and the strong force.[4]

As judged by experimental observation, GR is a perfectly successful theory; it has perfectly matched every experimental test of it to very great accuracy.

The first postulate of GR:
GR is a theory based on three postulates. The first postulate of GR is the equivalence principle which states that:

> *Equivalence Principle:* An observer in free-fall in a uniform gravitational field is in an inertial reference frame. (An observer stood upon the surface of the Earth is in an accelerated reference frame.)

---

[4] The floating spaces are associated with the Lie groups $U(1)$, $SU(2)$, $SU(3)$.

There is an often unstated assumption within both special relativity and GR that inertial reference frames are the fundamental reference frames of the universe. Special relativity is based on the idea that observers moving at constant velocities relative to each other are each in an inertial reference frame. GR extends this idea to include observers in free-fall in a uniform gravitational field; such observers are also in inertial reference frames.

A consequence of observers in free-fall being in an inertial reference frame is that observers who are not in free-fall within a gravitational field are not in an inertial reference frame. Thus we have the view that a gravitational field is equivalent to an accelerated reference frame. Hence the 'equivalence' in equivalence principle.

Clearly, the reference frame arbitrarily chosen by a mortal observer ought not to affect the physics of the universe. Using this idea, from the equivalence principle, we can deduce the gravitational bending of light rays, gravitational red-shift and gravitational time dilation. If we take the view that light moves in a straight line, as it does in inertial reference frames, then, from the gravitational bending of light rays, we can deduce that space-time is curved.

We can view the equivalence principle part of the theory of GR as being simply a statement of the consequences of a self-evident truth. If the self-evident truth is true, we can be certain that the consequences of it are true. Of course, self-evident truths are sometimes not true.

The second postulate of GR:
The second postulate of GR is the field equations which describe how space-time curves in response to mass-energy. GR associates gravity with the curvature of space-time. In the modern view, GR does not postulate that our 4-dimensional space-time is embedded in a curved way within a higher dimensional flat space. In the modern view, GR takes the curvature of space-time to be intrinsic curvature such as, for example, the curvature of a spherical surface which cannot be flattened. An intrinsically curved space contains its own geometry independently of any higher dimensional space in which the space sits just as the surface of a sphere contains its own

geometry independently of any higher dimensional space in which the sphere sits.

The GR field equations cannot be deduced from the equivalence principle; they are a separate postulate of the theory. Furthermore, the Riemann mathematical structure of the curved space-time within which the field equations are expressed is assumed by the theory rather than deduced.

We can view this second part of GR as a model of gravity constructed by humankind from thin air; how good the model is we determine by comparison with reality. We can never be certain that a model is true; a model is no more than very useful.

The third postulate of GR:
The third postulate of GR is the existence of the energy-momentum tensor. This is to say that the 'cause' of gravity is not a scalar field of simply mass but is a tensor field whose components are energy and momentum. This reflects the equivalence of mass and energy as expressed by the famous equation $E = mc^2$ of special relativity. Of course, Isaac Newton saw gravity as being 'caused' by only the mass of a body.

Within GR, we must accept that energy gravitates. An example of this is the pressure energy within a collapsing supernova. As the supernova shrinks in size the internal pressure increases and so the energy stored in that pressure increases and so the gravitational force associated with that energy increases and so the star is squeezed further and the pressure increases. The mathematics is such that eventually the gravitational force 'caused' by the pressure is so great that it overcomes the pressure force and so a black hole is formed.

What GR does not explain:
Although GR is perfectly successful as a theory that allows us to predict and understand phenomena, the structure of general relativity is not perfect in at least five ways:

a) The field equations of GR have to be guessed[5]; they cannot be deduced within the theory.

b) GR must assume that space-time is 4-dimensional; this cannot be deduced from the theory.

c) GR must assume the distance function of our space-time; this cannot be deduced from the theory.

d) GR assumes that space-time is of a Riemann nature including local flatness and the presence of two types of 2-dimensional angles and the associated rotations and inner products.

e) GR assumes the existence of the energy-momentum tensor.

Further, although GR encompasses and includes the special theory of relativity, GR stands separate from quantum field theory, QFT. The mathematical structure of GR does not allow gravity to be quantitised like classical electromagnetism is quantitised, and there is no sign of Planck's constant in traditional[6] GR. For anything other than microscopic physics, this is not a problem, but it is expected that we will need a quantum theory of gravity to deal with very intense microscopic gravitational phenomena.

Our inability to deduce the field equations of GR means that field equations other than the GR field equations might be the correct field equations rather than the ones we have guessed. Of course, any set of field equations must be in accord with observation, but that does not eliminate all possible proposed field equations. Known to your author, there are two other proposed sets of field equations which are in accord with observation. These are the Brans-Dicke field equations of the scalar tensor theory of gravity and the field equations of Elie Cartan's torsion theory of gravity. We prefer the Einstein field equations for no reason other than they are simpler than the other two proposals, but we cannot eliminate the other two sets of proposed field equations by observation.

---

[5] Einstein made three guesses. The first two guesses did not fit with observation; the third guess did fit with observation, and so we have his third guess for the field equations of GR.

[6] Although we have shown elsewhere that Planck's constant is within the complex numbers, $\mathbb{C}$, and that 2-dimensional Euclidean rotation is from the complex plane, this is not the traditional view.

GR gravity compared to the other forces of nature:
The GR theory of the gravitational force differs from the other forces of nature in that it is a force within space-time rather than over space-time. Given the observation that the gravitational force provides the same acceleration to different masses at the same location, we might have suspected that gravity could not be over space-time. Attempts have been made to write gravitation as a scalar field, a vector field, or a tensor field over space-time, but these attempts fail to match observation.[7] We are thus quite certain that the gravitational force is not over space-time. However GR does not explain why gravity is within space-time whereas the other forces of physics are over space-time.[8]

The mathematical structure of GR space-time:
Riemann geometry is the mathematics that underpins the GR field equations and the concept of intrinsic curvature. Riemann geometry was developed by Bernhard Riemann (1826-1866) in 1854; it is not derivable from the equivalence principle.

The Riemann nature of the space-time of GR is a construction of several parts and several assumptions.

a) A manifold which is a set of n-tuples of real numbers.
b) Local flatness of that manifold.
c) A type of flat space to be used as a tangent space to the manifold. There are different types of flat space.
d) An affine connection; a definition of the parallel transport of a vector.
e) A metric tensor; a measure of distance and angle.
f) Two types of 2-dimensional angles and the associated inner products. These are the Lorentz boost angles in 2-dimensional space-time and the Euclidean angles of a 2-dimensional flat plane.

The Riemann curvature tensor is constructed from the metric tensor by differentiation. The Ricci tensor and the Ricci scalar are formed by contracting the Riemann tensor. The Einstein tensor is

---

[7] See: John A Peacock - Cosmological Physics pg 27
[8] We will offer an explanation later.

constructed by subtracting half the product of the metric tensor with the Ricci scalar from the Ricci tensor. We see that the Einstein tensor is derived from the metric tensor. GR also assumes the existence of the energy-momentum tensor. Putting, by guesswork[9], the Einstein tensor equal to the energy-momentum tensor produces the field equations of GR.

The existence of locally flat manifolds, flat tangent spaces, affine connections, metric tensors, and 2-dimensional angles are assumptions of Riemann geometry. We will look at each of these components in more detail later when we offer explanations of why they exist and of how Riemann geometry emerges from the spinor algebras.

Summary:

Riemann space-time is based on assuming:

a) A manifold
b) Local flatness of the manifold
c) Flat tangent spaces
d) An affine connection
e) A metric tensor
f) 2-dimensional angles

GR is based on:

a) The equivalence principle
b) The GR field equations written in Riemann space-time
c) The energy-momentum tensor

---

[9] The guesswork is educated. These are the only two tensors we have that have zero divergence – which is how Einstein guessed their equality.

Chapter 3

The Genesis of Riemann Space

We will frequently use the super-imposition operation. The operation of super-imposition is no more than a generalisation of taking the expectation value of a set of quantum systems. It is by super-imposition that we move from the quantum universe of the spinor algebras to the classical universe of GR and classical electromagnetism.

The six A_3 algebras:

We assert that our 4-dimensional space-time emerges from the super-imposition of the six A_3 spinor algebra spaces as the emergent expectation space of those algebras.

The A_3 algebras are six of the eight non-commutative 4-dimensional double-cover spinor algebras within the finite group $C_2 \times C_2$. The other two non-commutative double-cover spinor algebras within the $C_2 \times C_2$ group are the left-chiral quaternions and the right-chiral quaternions. Each A_3 algebra has one real variable and three imaginary variables. The three imaginary variables are two square roots of plus unity and one square root of minus unity. The A_3 algebras are algebraically isomorphic. There are six of them because this one type of spinor algebra occurs in six different bases[10]. Less technically, there are six ways that two symmetric variables and one anti-symmetric variable can be combined to form a 4×4 matrix of multiplicatively closed form.

---

[10] See: Dennis Morris - The Physics of Empty Space

Two of the six A_3 algebras are the traditional SSA algebras[11]:

$$SSA = \exp\left(\begin{bmatrix} a & b & c & d \\ b & a & d & c \\ c & -d & a & -b \\ -d & c & -b & a \end{bmatrix}\right) \qquad SSA_{Anti} = \exp\left(\begin{bmatrix} a & b & c & d \\ b & a & -d & -c \\ c & d & a & b \\ -d & -c & b & a \end{bmatrix}\right)$$

$$(3.1)$$

For historical reasons, the two SSA algebras are named in the wrong order; the '*Anti*'[12] is in the wrong place. We correct that now, giving the six A_3 algebras as:

$$SSA^* = \exp\left(\begin{bmatrix} a & b & c & d \\ b & a & -d & -c \\ c & d & a & b \\ -d & -c & b & a \end{bmatrix}\right) \qquad SSA^*_{Anti} = \exp\left(\begin{bmatrix} a & b & c & d \\ b & a & d & c \\ c & -d & a & -b \\ -d & c & -b & a \end{bmatrix}\right)$$

$$(3.2)$$

$$SAS = \exp\left(\begin{bmatrix} a & b & c & d \\ b & a & d & c \\ -c & d & a & -b \\ d & -c & -b & a \end{bmatrix}\right) \qquad SAS_{Anti} = \exp\left(\begin{bmatrix} a & b & c & d \\ b & a & -d & -c \\ -c & -d & a & b \\ d & c & b & a \end{bmatrix}\right)$$

$$(3.3)$$

$$ASS = \exp\left(\begin{bmatrix} a & b & c & d \\ -b & a & -d & c \\ c & -d & a & -b \\ d & c & b & a \end{bmatrix}\right) \qquad ASS_{Anti} = \exp\left(\begin{bmatrix} a & b & c & d \\ -b & a & d & -c \\ c & d & a & b \\ d & -c & -b & a \end{bmatrix}\right)$$

$$(3.4)$$

With this renaming, the anti-symmetric variables of the algebras on the left together with a real variable form a left-chiral quaternion

---

[11] SSA stands for Symmetric, Symmetric, Anti-symmetric in the order of the variables $\{b, c, d\}$.

[12] Revision note: 'Anti' is really the wrong word. These algebras are either left-chiral or right-chiral.

while the anti-symmetric variables of the anti-algebras on the right together with a real variable form a right-chiral quaternion. Since emerging tensors like the electromagnetic tensor take the left-chiral quaternion form and 'drop' the right-chiral quaternion form, this renaming is essential clarification.

The A_3 algebras are the spinor algebra forms of the Clifford algebra $Cl_{2,0} \cong Cl_{1,1}$; these Clifford algebras are isomorphic as spinor algebras but are traditionally interpreted to be distinct.[13]

The emergent expectation space:
Super-imposition is just adding the isomorphic spinor algebras which derive from a particular finite group, but we cannot add spinor algebras written in different bases without destroying the algebraic structure. An essential part of that algebraic structure is the algebraic multiplication operation of the algebra.

Within a spinor algebra, multiplication has to have a particular set of properties; it has to satisfy the axioms of a division algebra[14]. One of these properties is multiplicative closure; a duck mated with a duck must produce a duck. Within a sum of algebras in different bases, this is impossible. (In classical physics we often see calculative procedures which are sloppily called multiplication. These procedures are often of the form of a camel mated with a donkey produces a chicken.)

With the multiplication operation destroyed, we cannot have rotation or inner products or any of the geometry within the algebra. In particular, we cannot have relations like $i^2 = -1$; we cannot have imaginary variables. However, the variables still exist. There are only two types of variables; these types are imaginary variables and real variables. If the variables cannot be imaginary, they must be real. Super-imposition destroys the algebra but leaves behind the variables as an infinite set of n-tuples of real numbers. A set of n-tuples of real numbers is a real manifold. This is a set of co-ordinate

---

[13] The interpretation distinguishes vectors from bi-vectors; the spinor algebra does not make this distinction but sees both vectors and bi-vectors as imaginary variables.

[14] Division algebra, spinor algebra – same thing.

points upon which we may place any arbitrary co-ordinate system we choose. It is because of the algebraic destruction that we do not see the spinor rotations, in the A_3 case, 4-dimensional rotations, in our 4-dimensional space-time.

All spinor spaces are flat. Any attempt to introduce intrinsic curvature into a spinor space destroys the algebra. Adding the six A_3 algebras destroys the algebraic spatial structure, but, at an infinitesimally small point, the difference of bases is irrelevant. At an infinitesimally small point, the six flat spaces of the six spinor algebras sit comfortably on top of each other and the flatness property of the spinor algebras is preserved through super-imposition. In Riemann geometry, we call this local flatness. The local flatness of the manifold is inherited from the flatness of the underlying spinor algebras. This does not mean the emergent manifold is globally flat. Without the algebraic structure, there is no necessity for the space to be globally flat. There is no algebraic structure to hold the space flat. The emergent manifold is flat at only infinitesimally small points. Every infinitesimally small point is flat, but not in the same 'flat plane' as other infinitesimally small points.

Curvature within a spinor algebra is meaningless, and so zero curvature is meaningless. The local flatness which has emerged from super-imposition is not zero curvature. We distinguish between local flatness and zero curvature. This is a quite important technical point because there are different types of curvature based on 2-dimensional rotations or based on 3-dimensional rotations or... or a type curvature fabricated from several spinor rotations. It is not necessarily the case that the local flatness space is of the same nature as the curvature of the space. Although technically important, there seems to be no physical consequences to this distinct from curvature nature of the local flatness.

When an emergent expectation space emerges from a super-imposition of isomorphic spinor algebras[15], it emerges as a locally flat real manifold. The dimension of the manifold is the dimension

---

[15] We reiterate, spinor algebras are types of complex numbers like the quaternions or the complex numbers. Spinor spaces are the same things as spinor algebras.

of the underlying spinor spaces because this is the number of variables in each spinor algebra.

The emergent manifold has dimension and local flatness but nothing more. There are no rotations within the emergent expectation space nor concepts such as angles, inner products, or vector curls. There is no concept of distance or of direction within the emergent expectation space. Because the A_3 algebras occur in six different bases, the symmetries, including rotation, inner products and vector curls, of the underlying A_3 algebras, are smashed by the act of super-imposition. (Of course, these symmetries remain intact within each of the separate spinor algebras and might correspond to quantum physics in the 'background' of the emergent expectation space, but that is not our present concern.)

Aside: This is reminiscent of the Higgs mechanism in which we break the symmetry by converting two complex numbers into four real numbers.

Tangent spaces:
Although a manifold inherits local flatness from the underlying spinor algebras, it does not inherit a particular type of flat space to be the tangent space at each point of the manifold. The manifold cannot inherit the flat space of the underlying spinor algebras because all the geometry of those algebras has been smashed by the super-imposition of those algebras.

There are different types of flat space. There are the flat spinor spaces such as the 2-dimensional flat Euclidean plane and the 2-dimensional flat space-time of special relativity or the four 3-dimensional flat spaces of the 3-dimensional spinor algebras. There are twenty-four types of flat 4-dimensional spinor space.

There are also flat spaces that are fabrications of a number of spinor spaces. We assert that many of the properties of our 4-dimensional space-time are properties of a fabrication of six 2-dimensional spinor spaces, three of each type. We assert that the flat tangent space of our 4-dimensional space-time is a fabrication of six copies of the 2-dimensional spinor spaces, three of each type. This is a type of flat space that is very different from an underlying flat A_3 space.

Many text-books illustrate a flat space tangent to a curved but locally flat manifold as a 2-dimensional flat Euclidean plane grazing the curved manifold at a single point.

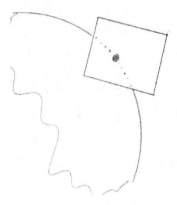

Although, when drawn properly, this is an edifying illustration, it hides the fact that the tangent space can be any type of flat space. There is no necessity what-so-ever for the flat tangent space to be Euclidean. In the absence of restraints or conditions, such as a distance function, the tangent space could be a space from any spinor algebra of the appropriate dimension, perhaps, say, a 4-dimensional spinor space from the C_4 group, or the tangent space could be a fabrication of more than one spinor space.

We are of the view that the only flat spaces are the spinor spaces and fabrications of those spinor spaces. We are of the view that the expectation distance function dictates the type of flat tangent space.

The expectation distance function:
An expectation distance function emerges alongside the emergent expectation manifold from the super-imposition of the distance functions of the underlying isomorphic spinor algebras. The distance function of a spinor algebra is the determinant of the matrix of the algebra. We usually use the Cartesian form of the spinor algebra for ease when we calculate the determinant.

The determinant of a spinor algebra is equal to d^n where n is the dimension of the spinor algebra; for the 4-dimensional A_3 algebras, this is d^4, and, for the 8-dimensional spinor algebras, this is d^8. We will see that emergent expectation spaces are fabricated from the 2-dimensional spinor spaces whose distance function is a quadratic form equal to d^2. The expectation distance function dictates the form of the fabricated emergent expectation space, but, to do this, it must be of the same form as the 2-dimensional spinor spaces, d^2. Hence, the expectation distance function of an emergent expectation space must be the expectation of the d^2 distance functions. This is achieved in the case of the A_3 algebras and the quaternions by simply taking the square root of the distance function thereby reducing d^4 to d^2. We then sum the d^2 distance functions of the isomorphic spinor algebras to form the expectation distance function of the emergent expectation space[16]. An A_3 distance function is of the form:

$$d^4 = \left(t^2 - x^2 - y^2 + z^2\right)^2$$
$$d^2 = t^2 - x^2 - y^2 + z^2$$

(3.5)

In the case of the A_3 algebras, using the Cartesian matrix form of these algebras, we have the expectation distance function:

$$SUM \begin{cases} dist^2 = t^2 - x^2 - y^2 + z^2 \\ dist^2 = t^2 - x^2 - y^2 + z^2 \\ dist^2 = t^2 - x^2 + y^2 - z^2 \\ dist^2 = t^2 - x^2 + y^2 - z^2 \\ dist^2 = t^2 + x^2 - y^2 - z^2 \\ dist^2 = t^2 + x^2 - y^2 - z^2 \end{cases} = 2\left(3t^2 - x^2 - y^2 - z^2\right) \quad (3.6)$$

---

[16] We take the root and then sum rather than first summing and then taking the root. We want the expectation d^2 distance function not the root of the d^n expectation distance function.

The 2 is just a scaling factor and the 3 is just the units in which we measure time or space. The reader will recognise this as the distance function of our 4-dimensional space-time.

The only other spinor algebras with a distance function which will reduce to d^2 form are the left-chiral quaternions and the right-chiral quaternions. The quaternion emergent expectation space has no time within it. The 8-dimensional $C_2 \times C_2 \times C_2$ spinor algebras (remember, these are the 8-dimensional Clifford algebras) have emergent expectation spaces that fall apart into 2-dimensional spaces, and so all the higher dimensional $C_2 \times C_2 \times C_2 \times...$ spinor algebras similarly must fall apart into 2-dimensional spaces. This effectively means that our 4-dimensional space-time is the unique emergent expectation space with time. (We think the quaternions are connected to the weak force.)

We now have a locally flat manifold with a distance function. We need a tangent space. We assert that the type of tangent space must fit with the expectation distance function that has emerged alongside the manifold. This severely constrains the type of flat tangent 'plane' which may be affixed to the manifold; for a start, in the case of the A_3 algebras, the tangent 'plane' will be 4-dimensional.

We conjecture that, in general, the expectation distance function will not be the same as any spinor space distance function – it is the sum of isomorphic spinor space distance functions. (We are very confident that this conjecture is correct, but we have no proof.) In our particular case, we can simply check all spinor space distance functions of the appropriate dimension. If the emergent flat tangent space is not one of the spinor spaces, it must be a fabrication of more than one of the spinor spaces or nothing.

The expectation distance function of the A_3 algebras is:

$$d^2 = t^2 - x^2 - y^2 - z^2 \qquad (3.7)$$

There is no 4-dimensional spinor space with this distance function, therefore, if this is to be a tangent space, it must be a fabrication of spinor spaces. None of the 3-dimensional spinor spaces have a distance function which fits any of the sets of three of the variables in (3.7). However, there are 2-dimensional spinor spaces, the complex plane, $\mathbb{C}$, and the hyperbolic complex numbers, $\mathbb{S}$, which equal (3.7) when any two of the variables in (3.7) are zero. Those 2-dimensional distance functions are:

$$d^2 = t^2 - a^2 \quad \& \quad d^2 = b^2 + c^2 \qquad (3.8)$$

With a little thought, we see that we can form a flat tangent space as a fabrication of three copies of each of the two 2-dimensional spaces - six 2-dimensional spaces altogether – by choosing any two variables from the four variables in (3.7). We conclude that the tangent space of our 4-dimensional space-time is a fabrication of six copies of 2-dimensional spinor spaces, three of each kind.

We allow that our 4-dimensional space-time is also a fabrication of four copies of the 1-dimensional spinor space which is the real numbers, but the 1-dimensional spinor space has no angles, curls, or anything geometrically interesting and we do not notice it hiding under the 2-dimensional spinor spaces.

Angles:
To fit our 4-dimensional space-time together properly, we need six 2-dimensional angles. There is one angle in each of the six A_3 spaces.

By coincidence, we have the needed number of angles to form our 4-dimensional space-time.

In the quaternion case, there are only two quaternion algebras and so only two available angles. The quaternion emergent expectation space is a 4-dimensional space in which one can rotate in only two orthogonal planes. These two rotations cannot be combined together to make a rotation in another plane because there are no more angles. The consequence of this is that rotation is directionally quantitised; you can have only spin up or spin down. This seems to be the space of electron spin.

Orthogonality:
There are two sources of the orthogonality of the axes of the tangent space of the emergent expectation space.

The first source is that the expectation distance function is formed from four independent real variables. The six 2-dimensional spaces are held in their orthogonal positions within this 4-dimensional tangent space of the emergent expectation space by the orthogonality of the variables in the distance function, but, as with the tangent space, this holding in position is only local. There is nothing to ensure the orientation of the set of axes of the space does not change from point to point in the emergent expectation space.

In Riemann geometry, this orthogonality corresponds to local flatness. Later, we will see that our 4-dimensional space-time emerges from the symmetric variables of the A_3 algebras. (The anti-symmetric variables lead to classical electromagnetism.)

The second source is that orthogonality is 'forced' on to the tangent space by the mathematical fact that any symmetric matrix with real elements can be made diagonal by a suitable choice of co-ordinate system. To make this 'forcing' work, we will have to extract the anti-symmetric variables from the A_3 algebras - see later. Such extraction of the anti-symmetric variables does not affect the orthogonality of the remaining variables; there are six A_3 algebras; that is enough symmetric variables to go around four times.

Aside: The distance function of our space-time above, (3.7), is written in Cartesian co-ordinates because we used the Cartesian matrix form of the A_3 algebras to calculate it. Spinor algebras are flat, but they are not constrained to be written in only Cartesian co-ordinates. Spinor algebras are most easily written in their polar forms. In this case, the expectation distance functions would be written in non-Cartesian co-ordinates.

Super-imposition of 2-dimensional spinor algebras:
Hang on! the attentive reader cries. Surely, we should use the super-impositions of the two 2-dimensional algebras rather than the raw algebras in the fabrication of our 4-dimensional space-time– good point, well spotted!

The two 2-dimensional spinor algebras, the Euclidean complex numbers, $\mathbb{C}$, and the hyperbolic complex numbers (2-dimensional space-time), $\mathbb{S}$, derive from the finite group C_2. The derivation produces only one copy of each algebra[17]. If we super-impose a single algebra then nothing changes. In the cases of the 2-dimensional spinor algebras, the whole of the algebraic structure survives super-imposition. This means that 2-dimensional Euclidean rotations and Lorentz boosts (space-time rotations) are not destroyed by taking expectation values and other similar expectation objects. Indeed, the whole of the 2-dimensional algebraic structure including 2-dimensional angles, 2-dimensional curls, and 2-dimensional inner products survives super-imposition. The 2-dimensional trigonometric functions $\{\cosh(\ ),\sinh(\ )\}$ and $\{\cos(\ ),\sin(\ )\}$ also survive super-imposition. The survival of the 2-dimensional inner products are of importance because we can use the 2-dimensional inner products to fabricate a 4-dimensional inner product. From the 4-dimensional inner product, we get a metric tensor.

The 4-dimensional inner product:
The two 2-dimensional spinor algebras contain the two inner products[18]:

$$\vec{X} \cdot \vec{Y} = x^1 y^1 + x^2 y^2 \quad \& \quad \vec{X} \cdot \vec{Y} = x^1 y^1 - x^2 y^2 \qquad (3.9)$$

We assert that, constrained by the form of the expectation distance function and by the local orthogonality of the variables in the expectation distance function, these two 2-dimensional inner products are fabricated into the 4-dimensional inner product of what we call 4-vectors. That 4-dimensional inner product is:

$$\vec{X} \cdot \vec{Y} = x^0 y^0 - x^1 y^1 - x^2 y^2 - x^3 y^3 \qquad (3.10)$$

We do not assert that the form of this inner product is universally the same throughout the whole extent of the emergent expectation space. We assert that this is a local form only because the orthogonality is local only.

---

[17] This is because there is only one scaling parameter in 2-dimensions. See: Dennis Morris - The Physics of Empty Space.
[18] See: Dennis Morris Complex Numbers The Higher Dimensional Forms.

Aside: We often see 4-vectors mentioned within texts upon 2-dimensional special relativity. 4-vectors have no place in special relativity, and they do not work properly (we have to fudge the sign under a square root of the acceleration 4-vector). Special relativity is concerned with linear transformations within a spinor algebra space. There are no fabricated inner products within a spinor algebra.[19]

Conventionally, there is no such thing as the cross product of two 4-vectors. Within Clifford algebra, a cross product of two 4-dimensional vectors is a 2-dimensional plane rather than a vector. Perhaps we could fabricate a 4-vector cross product, but it would do no more than measure the angle between two vectors as the inner product does.

A fibre bundle is the affine connection:
Each individual A_3 algebra is rigidly flat. This means the four variables within it are all rigidly independent of each other. This is perhaps most easily, though wrongly, visualised by saying the axes of an A_3 space are all rigidly at right-angles to each other. The nature of each of the A_3 spaces is that it contains three 2-dimensional spaces because the $C_2 \times C_2$ group contains three C_2 sub-groups. These are double cover spinor 2-dimensional spaces; they are 4-dimensional 2-dimensional spaces but the sub-space structure is of a 2-dimensional nature.

We assert that the emergent expectation space, our 4-dimensional space-time, is the underlying space of a fibre bundle. We assert that, at each point of our 4-dimensional space-time, there is fixed an A_3 spinor space. We can visualise the A_3 space being pinned to our underlying space-time through the origin of the A_3 space at each point of our space-time. No, let us pin all six A_3 spaces through their origins to the underlying 4-dimensional space-time at each point in that underlying 4-dimensional space-time. Our fibre bundle is now

---

[19] See : Dennis Morris : Empty Space is Amazing Stuff

comprised of one underlying emergent expectation space and six A_3 spaces.

At each point, in the underlying space-time, the distance functions of the A_3 algebras are added to form the expectation distance function of the underlying space-time. Similarly, at each point, in the underlying space-time, the 4-dimensional orientations of the A_3 algebras are added to form an expectation orientation of the tangent space with respect to of the underlying space-time.

There is no constraint upon the orientations of the six A_3 algebras with respect to the underlying emergent expectation space. With nothing to fix it firmly, the expectation orientation of the tangent space floats freely.

We present the above again in a different way. Within each A_3 algebra, there is a 4-dimensional angle. The sum of these six 4-dimensional angles is just the sum of six triples of real numbers, and so there is a single 4-dimensional expectation angle associated with the underlying emergent expectation space, our 4-dimensional space-time, at every point in that emergent expectation space. We assert that this expectation angle varies continuously from point to point in our 4-dimensional space-time. We are saying the orientation of the tangent space, whose nature is dictated by the expectation distance function, with respect to the underlying emergent expectation space varies from point to point in the emergent expectation space.

This local variation of angle, or phase as it is most often called, is not a novel idea. Indeed, it is central to the whole of QFT wherein the electromagnetic, the weak and the strong forces emerge from the mathematics by exactly this local variation of angle within spaces attached as a fibre bundle to each point in our underlying space-time. In QFT, the attached spaces are called Lie groups; they are the Lie groups $U(1)$, $SU(2)$, and $SU(3)$ respectively.

We would substitute spinor algebras $\mathbb{C}$, $\mathbb{H}$, and an 8-dimensional Clifford algebra for the Lie groups and we would include the hyperbolic complex numbers, which gives us the $F = ma$ force, and we include the A_3 algebras as we have above.

Hang on! The attentive reader cries again. It is meaningless to say that the expectation orientation angle varies from point to point over a manifold in which there is no sense of direction. The attentive reader is correct, and this is a very important point.

We assert that the local variation of expectation orientation angle induces into the emergent expectation manifold an affine connection; we say the local variation of expectation orientation angle induces into the emergent expectation manifold a sense of direction and, with it, a notion of the parallel transport of a vector. In short, we are inducing intrinsic curvature into the emergent manifold. (Affine connections and intrinsic curvature will be discussed in more detail later.) Furthermore, the induced affine connection must be compatible with the expectation distance function; it must be the Levi-Civita connection.

I'll say that again a little differently. Imagine a vector aiming in a given direction at a given point in the emergent expectation space. Imagine another vector at a different point in the emergent expectation space. Are the two vectors parallel? Without the concept of direction in the emergent expectation space, we cannot compare the directions of two spatially separated vectors. What we can do, is to declare that all A_3 vectors are parallel. This induces into the emergent expectation space an affine connection, a set of parallel lines. This affine connection is intrinsic curvature within the emergent expectation space.

If the gravitational force is associated with the curvature of our 4-dimensional space-time, then, as with the forces described by QFT, we have the gravitational force emerging from a local variation of phase of a spinor algebra.

Aside: Taken together, the A_3 algebras double cover the Lie group called the Lorentz group, $SO(3,1)$.

Once we have an affine connection in the space, it makes sense to speak of the angle between two vectors in that space, and the inner product of two vectors makes sense.

Since the local orthogonality of the variables in the expectation distance function fixes the local orthogonality of the variables in the fabricated 4-dimensional inner product of the tangent space, the fabricated inner product will vary continuously from point to point within the emergent expectation space.

The metric tensor:

As we will see later, the metric tensor of a Riemann space is no more than inner products. It is a generalised measure of the length of a vector and of the angle between two vectors. If the fabricated inner product varies from point to point, the metric will vary from point to point in space-time. We call this variation of the metric from point to point intrinsic curvature.

To have curvature, we need to be able to measure change of orientation of a vector as it is transported along a path in the space. To measure change of orientation, we need a type of angle. Different spinor spaces have different types of rotation and different types of angle. Thus, flat tangent spaces fabricated from different spinor spaces will, as they are moved from point to point in the space, be associated with different types of curvature.

In our 4-dimensional space-time, we measure change of orientation with 2-dimensional angles and a 4-dimensional inner product that is a fabrication of 2-dimensional inner products. This curvature is a 4-dimensional curvature fabricated from 2-dimensional curvatures.

A counter example might clarify this. There are six single real numbers associated with the intrinsic curvature of a 4-dimensional Riemann space. These are called the principle curvatures. The principle curvatures correspond to the six possible 2-dimensional rotations in 4-dimensional Riemann space. If, instead of 2-dimensional rotations, the tangent space was fabricated from the 3-dimensional spinor spaces with the 3-dimensional rotations from the C_3 group, then, since one 3-dimensional rotation involves three axes, we would have four principle curvatures because there are four ways to choose three axes from four axes. Each of these 3-dimensional principle curvatures would be two real numbers because the nature of 3-dimensional spinor rotations is that the trigonometric functions take two arguments.

Summary:

We have seen a manifold with dimension and local flatness emerge from the super-imposition of the A_3 algebra. We have seen an expectation distance function emerge from the super-imposition of the distance functions of the A_3 algebras. We have seen how the expectation distance function and the dimension of the manifold dictates the nature of the tangent space as a fabrication of spinor spaces. We have seen that the spinor spaces in the fabrication bring with them their own types of rotation, trigonometric functions, and inner product. We have seen that the inner products are fabricated together to form the inner product of emergent expectation space. We have seen how the locally varying phase of an A_3 algebra induces an affine connection into the emergent expectation space. We have seen how the Riemann space we call space-time has emerged from the A_3 algebras. We now have what we need to build Riemann geometry and GR.

Important note:

Although, it was not known at the time this book was written. The $C_2 \times C_2$ finite group is the only finite group (there are an infinite number of finite groups) which produces an emergent expectation space that can support rotations[20]. Thus, our 4-dimensional space-time is the only geometric space that exists, except for the quaternion space which seems to be the space of the electro-weak force.

---

[20] See : Dennis Morris : The Uniqueness of our Space-time

Chapter 4

The Mathematics of Riemann Space I

We now begin to construct the mathematics of Riemann space. This is the mathematics in which the field equations of GR are written. Riemann geometry was originally constructed by Riemann building upon the work of Gauss.

Co-ordinate Transformations:
Clearly, the physics of the universe cannot vary with humankind's arbitrary choice of co-ordinate system. A n-dimensional set of co-ordinate points is an infinite set of n-tuples of real numbers.[21] This is called a manifold. There is no sense of distance between points or sense of direction or sense of rotation in a manifold, but we assume that the space is locally flat.

Our immediate task is to understand how the different co-ordinate systems we might impose upon a set of co-ordinate points are related to each other; how we can change from one co-ordinate system to another. Luckily for us, this is well understood, and we can stand upon the shoulders of our forebears in this regard; perhaps we should sit at their feet. We follow the standard mantra in this matter.

Aside: The reader might be familiar with the idea of changing variables as a way of simplifying a differential equation. Such change of variables is no more than a co-ordinate transformation.

Given a particular co-ordinate system, at each co-ordinate point, aiming in the directions of each of the co-ordinate axes, we can place basis vectors each tangent to each axis which we denote by $\vec{e}_i$. Notice that we put the index of a basis vector as a subscript. This is part of

---

[21] We have no interest at all in co-ordinates which are formed from some other type of number such as the complex numbers, $\mathbb{C}$.

the notation of tensor calculus. All basis vectors have subscripted indices. We have:

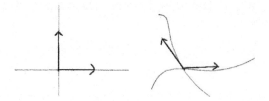

Notice there is no concept of orthogonality (right-angles) in a manifold, but there is independence of each co-ordinate axis.

These basis vectors, $\vec{e}_i$, we declare to be of unit length. This is not the imposition of a measure of distance between the co-ordinate points; it is merely that every whole real number is, an unscaled, one unit apart from its immediate neighbours in the real number line. If we want, we can say that these basis vectors follow the direction of the co-ordinate axes, which might writhe around in all sorts of directions; we can get away with saying this because the concept of direction is not within the co-ordinate system and so no-one can say otherwise. Our statement about direction is meaningless really, but it is a mental crutch upon which to lean.

We will begin by considering a simple change of co-ordinate system within the 2-dimensional Euclidean plane. The 2-dimensional Euclidean plane is a fully formed space, not just a manifold. The ideas we use the Euclidean plane to illustrate apply to a manifold, but it is easier to see these ideas in a fully formed space. The simple change of co-ordinate system we choose is just a rotation of the axes.

We will begin with our axes horizontal and vertical and we will consider a vector at the origin. This is a flat space; this has nothing to do with curvature. We have:

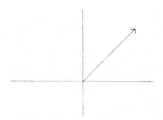

We will now change the co-ordinate system by rotating the co-ordinate axes through the angle θ. As we do this, because the basis vectors aim along the co-ordinate axes, the basis vectors will change direction. We have:

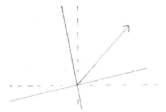

The change in the basis vectors is given by:

$$\begin{bmatrix} \vec{e}_x' \\ \vec{e}_y' \end{bmatrix} = \begin{bmatrix} \cos\theta & \sin\theta \\ -\sin\theta & \cos\theta \end{bmatrix} \begin{bmatrix} \vec{e}_x \\ \vec{e}_y \end{bmatrix} \tag{4.1}$$

Now, the vector, which represents something real, is unchanged in direction or length by our arbitrary choice of co-ordinate system, but the vector's components will be different in our new co-ordinate system. How do the components of the vector change to accommodate our change of co-ordinate system? They must rotate in the opposite direction to the co-ordinate axes. We have:

$$\begin{bmatrix} V^{x'} \\ V^{y'} \end{bmatrix} = \begin{bmatrix} \cos(-\theta) & \sin(-\theta) \\ -\sin(-\theta) & \cos(-\theta) \end{bmatrix} \begin{bmatrix} V^{x} \\ V^{y} \end{bmatrix} \tag{4.2}$$

We see that, under this change of co-ordinate system, the components of the vector change in the opposite direction to the change in the basis vectors. We will say that again in case the reader is not as alert as she ought to be; under this change of co-ordinate system, the components of the vector change in the opposite direction to the change in the basis vectors. Although we have shown this using a very simple change of co-ordinate system, this is true for any change of co-ordinate system. We say that the basis vectors change covariantly and the components of the vector change contravariantly[22]. If we are to use co-ordinate transformations, and they are all over the place in general relativity, we need to keep track of the fact that basis vectors transform differently from the components of vectors. This is done remarkably well by the 'index notation' of Riemann geometry.

Index notation:

We write objects, like the components of vectors, which transform contravariantly with an upper index, superscript, V^x. (Our memory of this is aided by noticing that there is an important 't' in contravariant and an important 't' in top.) We write objects, like basis vectors which transform covariantly with a lower index, $\overrightarrow{e}_x$. (We notice that there is not an important 't' in covariant.) There is nothing fundamentally special about choosing lower indices for basis vectors; it is arbitrary whether humankind puts the indices as subscripts or superscripts; it is purely notation, but the convention is as shown above, and no-one differs from this convention.

It is not stretching the imagination too far to expect that more complicated co-ordinate transformations from one type of co-ordinate system to another type of co-ordinate system might also have objects which transform contravariantly (do you see the important 't'? does it remind you of 'top'?) and objects which transform covariantly.

Covariantly transforming vectors:

Looking back at our change of co-ordinates by rotation (4.1) & (4.2). How do the components of the vector change if we rotate the co-ordinates in the opposite direction? They change oppositely to the way they changed before. By reversing the direction of rotation of the co-ordinate axes, we reverse the direction in which the components of the vector change, and so we will need vectors whose components transform covariantly rather than contravariantly.

---

[22] Within a spinor algebra space, covariant and contravariant correspond to the two sides of the conjugation operation. Conjugation is just reverse rotation.

In a spinor space, like the complex plane, $\mathbb{C}$, an inner product is just a rotation and its reverse multiplied together to produce a number on the positive half of the real axis. For example:

$$\begin{bmatrix} \cos\theta & \sin\theta \\ -\sin\theta & \cos\theta \end{bmatrix} \begin{bmatrix} \cos\theta & -\sin\theta \\ \sin\theta & \cos\theta \end{bmatrix} = \begin{bmatrix} 1 & 0 \\ 0 & 1 \end{bmatrix} \tag{4.3}$$

Within Riemann geometry, the inner product (dot-product) of two vectors is written as $V_\mu W^\mu$. The same index in upper and lower positions indicates summation. If the inner product of two vectors:

$$V_0 W^0 + V_1 W^1 + V_2 W^2 + V_3 W^3 = V_\mu W^\mu = s \tag{4.4}$$

is to be invariant under a co-ordinate transformation, and it is only a real number, a scalar, and so it must be invariant, then the components V_μ must transform oppositely to the components W^μ.

It is perhaps surprising to discover that there are some 'vectors' whose components transform covariantly; we call these covariant vectors, or 1-forms, as opposed to 'normal' vectors which we call contravariant vectors. The gradient of a scalar field is such a covariant vector.

The gradient of a scalar field, S, is:

$$\nabla S = \begin{bmatrix} \dfrac{\partial S}{\partial x^1} & \dfrac{\partial S}{\partial x^2} & \dfrac{\partial S}{\partial x^3} & \cdots \end{bmatrix} \tag{4.5}$$

The reader is advised that, in general relativity, we refer to different co-ordinates as x^i rather than as $\{x, y, z\}$. In tensor calculus, we take the view that $\dfrac{\partial}{\partial x^i}$ is really a set of basis vectors, and so the components of the gradient, (4.5), are really basis vectors, and so we might expect them to transform covariantly under change of co-ordinate system; they do. We say that the gradient is a covariant vector. Technically, for a given scalar field, ϕ, members of the set $\dfrac{\partial \phi}{\partial x^i}$ are components of covariant vectors (also called 1-forms).

Another way to think of this is to think the 'bits' of a vector are $\overrightarrow{ae_i}$; this is two 'things' multiplied together; if we can choose one of these 'things' to be a basis vector, then we could choose the other and write $\overrightarrow{a_i e}$; the 'length' of the a is just a real number which can be transferred into the e so that the length of the a is now unity. We can bring about the change by differentiating the scalar field, a. Differentiating a scalar field produces a vector (the gradient).

Vectors that are tangent to a curve through space specified by a real parameter, λ, are contravariant vectors. By combining contravariant vectors with contravariant vectors, see (5.2), or contravariant vectors with covariant vectors or covariant vectors with covariant vectors we can build tensors. (Tensors are just products of vectors - see later.)

Aside: Scalars are sometimes called 0-forms. Doubly covariant tensors, $T_{\alpha\beta}$, are sometimes called 2-forms. Tensors of the triply covariant form $T_{\alpha\beta\chi}$ are sometimes called 3-forms.

Note: Tensors, of which vectors are one example, come in different types defined by how they transform under a change of co-ordinate system; that is how their components change with a change of co-ordinate system. If we differentiate a tensor which transforms entirely contravariantly, it becomes a tensor which, in one of its indices, transforms covariantly. In some ways, the difference between contravariant and covariant is differentiation. Rotation is like differentiation; the trigonometric functions change in the same way under rotation (by 90°) as they do under differentiation.

The covariant nature of some vector components, like the gradient, will become clearer shortly; do not be daunted; read on.

Let's do it again:

A change, transformation to use the posh terminology, in co-ordinate system is accomplished with a square transformation matrix, $\Lambda_{b'}^{a}$. The size of the transformation matrix is the same as the dimension of the space. The elements of the transformation matrix are given, in the case of the basis vectors, by (there is an example soon):

$$\Lambda^a_{b'} = \frac{\partial x^a}{\partial x^{b'}}$$

$$\vec{e_a} = \Lambda^{b'}_a \vec{e_{b'}}$$

(4.6)

Note that, in the lower of these expressions, we have to sum over the values of b'; this notation is known as the Einstein summation convention, more on this later; for now, see the example immediately below, (4.10). The components of the transformation matrix are the ratios of a change in one old co-ordinate to the change in one new co-ordinate.

Aside: The transformation matrix is sometimes called the Jacobean matrix, and its determinant is sometimes called the Jacobean.

We give an example. In flat 2-dimensional Euclidean space, the Cartesian co-ordinates, (x, y), are related to the polar co-ordinates, (r, θ), by:

$$x = r \cos \theta \quad : \quad y = r \sin \theta$$

(4.7)

From these we get the transformation matrix of the basis vectors as:

$$\Lambda^x_r = \frac{\partial x}{\partial r} = \cos \theta \qquad \Lambda^y_r = \frac{\partial y}{\partial r} = \sin \theta$$

$$\Lambda^x_\theta = \frac{\partial x}{\partial \theta} = -r \sin \theta \qquad \Lambda^x_r = \frac{\partial y}{\partial \theta} = r \cos \theta$$

(4.8)

We then have the transformation of the basis vectors:

$$\begin{bmatrix} \vec{e_r} \\ \vec{e_\theta} \end{bmatrix} = \begin{bmatrix} \cos \theta & \sin \theta \\ -r \sin \theta & r \cos \theta \end{bmatrix} \begin{bmatrix} \vec{e_x} \\ \vec{e_y} \end{bmatrix}$$

(4.9)

This can be expressed otherwise as:

$$\vec{e_r} = \cos \theta \vec{e_x} + \sin \theta \vec{e_y}$$

$$\vec{e_\theta} = -r \sin \theta \vec{e_x} + r \cos \theta \vec{e_y}$$

(4.10)

We see the new basis vectors are a 'mixture' of the old basis vectors.

Note:

$$\vec{e_r} \cdot \vec{e_r} = \left(\cos\theta \vec{e_x} + \sin\theta \vec{e_y}\right)\left(\cos\theta \vec{e_x} + \sin\theta \vec{e_y}\right)$$

$$= \left(\cos^2\theta\right)\vec{e_x} \cdot \vec{e_x} + \left(\cos\theta\sin\theta\right)\vec{e_x} \cdot \vec{e_y}$$

$$+ \left(\cos\theta\sin\theta\right)\vec{e_y} \cdot \vec{e_x} + \left(\sin^2\theta\right)\vec{e_y} \cdot \vec{e_y} \qquad (4.11)$$

$$= \cos^2\theta + \sin^2\theta = 1$$

And similarly:

$$\vec{e_\theta} \cdot \vec{e_\theta} = r^2 \qquad (4.12)$$

We see that the inner product of a basis vector with itself is not always unity in polar co-ordinates. This is because the length of the basis vector concerned is not constant in polar co-ordinates.

The inner products of basis vectors with themselves or other basis vectors varies from one co-ordinate system to another. This is an important point; we will in due course define the components of a very important tensor called the metric tensor to be such inner products of basis vectors. Thus the metric tensor will vary with the co-ordinate system.

Of course, just as we transformed in the opposite direction above by rotating in the opposite direction, (4.2), we could transform our example in the opposite direction starting from:

$$r = \sqrt{x^2 + y^2} \quad : \quad \theta = \tan^{-1}\left(\frac{y}{x}\right) \qquad (4.13)$$

$$\Lambda^r_x = \frac{\partial r}{\partial x} = \frac{x}{\sqrt{x^2 + y^2}} \qquad \Lambda^r_y = \frac{\partial r}{\partial y} = \frac{y}{\sqrt{x^2 + y^2}}$$

$$\qquad (4.14)$$

$$\Lambda^\theta_x = \frac{\partial \theta}{\partial x} = \frac{x}{x^2 + y} \qquad \Lambda^\theta_y = \frac{\partial \theta}{\partial y} = -\frac{y}{x^2 + y}$$

Instead of transforming basis vectors, this will transform the components of the vectors:

$$\begin{bmatrix} V^r \\ V^\theta \end{bmatrix} = \begin{bmatrix} \dfrac{x}{\sqrt{x^2+y^2}} & \dfrac{y}{\sqrt{x^2+y^2}} \\ \dfrac{x}{x^2+y} & -\dfrac{y}{x^2+y} \end{bmatrix} \begin{bmatrix} V^x \\ V^y \end{bmatrix} \qquad (4.15)$$

If the vector we were transforming was a covariant vector such as the gradient, then its components would transform oppositely. It is conventional to take the view that the 'basis vectors' of a covariant vector are different from the basis vectors of a contravariant vector and to call them basis 1-forms which are often designated by ϖ^i. Basis 1-forms transform the same way as the components of a contravariant vector:

$$\begin{bmatrix} \varpi^r \\ \varpi^\theta \end{bmatrix} = \begin{bmatrix} \dfrac{x}{\sqrt{x^2+y^2}} & \dfrac{y}{\sqrt{x^2+y^2}} \\ \dfrac{x}{x^2+y} & -\dfrac{y}{x^2+y} \end{bmatrix} \begin{bmatrix} \varpi^x \\ \varpi^y \end{bmatrix} \qquad (4.16)$$

Clearly, the components of a covariant vector will transform oppositely to the basis 1-forms. We have:

$$\begin{bmatrix} V_r \\ V_\theta \end{bmatrix} = \begin{bmatrix} \cos\theta & \sin\theta \\ -r\sin\theta & r\cos\theta \end{bmatrix} \begin{bmatrix} V_x \\ V_y \end{bmatrix} \qquad (4.17)$$

The reader will see that we have written the contravariant vector components with upper indices (super-script) and have written the covariant vector components with lower indices (sub-script). This is ubiquitous basic notation within tensor calculus.

Contravariant vectors and co-variant vectors:
It would seem as if we have two types of vectors. The vectors whose components we designate with an upper index we call contravariant vectors. The vectors whose components we designate with a lower index we call covariant vectors[23]. They are both vectors, but the

---

[23] Covariant vectors are also known as: linear functions, linear functionals, dual vectors, bra-vectors, co-vectors and 1-forms.

notation is an extremely useful way of keeping track of whether we are transforming one way or the other way. The only difference between contravariant vectors and covariant vectors is the direction of the co-ordinate transformation of their components or of their basis vectors.

The transformation of each component or each basis vector is done by multiplying it by the different differentials and summing see (4.8) and (4.14). This is just multiplication by the transformation matrix, but we look at individual components. We have the contravariant case in 2-dimensions:

$$\left(V^m\right)' = \frac{\partial y^m}{\partial x^p}V^p = \frac{\partial y^m}{\partial x^1}V^1 + \frac{\partial y^m}{\partial x^2}V^2 \qquad (4.18)$$

Notice how we have summed over the p. The covariant case is:

$$\left(V_m\right)' = \frac{\partial x^m}{\partial y^p}V_m = \frac{\partial y^1}{\partial x^p}V_1 + \frac{\partial y^2}{\partial x^p}V_2 \qquad (4.19)$$

In these two expressions, (4.18) & (4.19), we have used the Einstein convention regarding summation, which we will cover again shortly. Notice that, taking the covariant vector in (4.19) to be the gradient, we have:

$$\left(\frac{\partial S}{\partial y^m}\right)' = \frac{\partial x^m}{\partial y^p}\frac{\partial S}{\partial x^m} = \frac{\partial S}{\partial y^p} \qquad (4.20)$$

This is why the gradient is a covariant vector.

Another way to think of the gradient is as an 'arrow' pointing uphill across a number of equipotential lines (height contours). The 'steepness' of the gradient is the number of equipotential lines the 'arrow' crosses per 'unit length'. The vectors tangent to the equipotential lines are the 'true' vectors called contravariant vectors; the 'arrows' that cross the equipotential lines are the 1-forms called covariant vectors. (The gradient can be a vector only if we have a distance function defining the 'unit length'. We need a metric (notion of length) to make the gradient into a vector. We will see in due course that the gradient can be made into a vector by 'raising its index' using the metric tensor.)

Looking at (4.19) & (4.20), we see that, in short, to swap from covariant co-ordinate transformations to contravariant co-ordinate transformations, we just invert the differentials. The difference is whether we are going from co-ordinate system A to co-ordinate system B or from co-ordinate system B to co-ordinate system A.

The Einstein convention and index slinging:

The Einstein convention is merely a notational device to indicate summation. It says that if a particular index appears in both upper and lower positions within an expression, then this means that terms in the expression are to be summed over the range of that index. We have:

$$V^s W_s = \sum_s V^s W_s = V^0 W_0 + V^1 W_1 + V^2 W_2 + \ldots \qquad (4.21)$$

Working with indices is often colloquially called 'index slinging'; it is like gun slinging but is less socially acceptable. There are several things to be careful about when 'index slinging'.

$$a_{ij}\left(x_i + y_j\right) \neq a_{ij} x_i + a_{ij} y_j$$

$$a_{ij} x_i y_j \neq a_{ij} y_i x_j \qquad (4.22)$$

$$\left(a_{ij} + a_{ji}\right) x_i y_j \neq 2 a_{ij} x_i y_j$$

Useful identities are:

$$a_{ij}\left(x_j + y_j\right) = a_{ij} x_j + a_{ij} y_j$$

$$a_{ij} x_i y_j = a_{ij} y_j x_i$$

$$a_{ij} x_i x_j = a_{ji} x_i x_j \qquad (4.23)$$

$$\left(a_{ij} + a_{ji}\right) x_i x_j = 2 a_{ij} x_i x_j$$

$$\left(a_{ij} - a_{ji}\right) x_i x_j = 0$$

Co-ordinate transformations in higher dimensions:

We have shown the 2-dimensional co-ordinate transformations above using the transformation matrices. Clearly in 4-dimensions, the transformation matrix would be a 4×4 transformation matrix containing differentials.

What has this to do with general relativity?:

Special relativity is about 2-dimensional rotations in 2-dimensional space-time. As such, a special relativity transformation is just a rotational change of co-ordinates. This co-ordinate transformation is done with a simple 2×2 rotation matrix:

$$\begin{bmatrix} \cosh \chi & \sinh \chi \\ \sinh \chi & \cosh \chi \end{bmatrix} \equiv \begin{bmatrix} \gamma & v\gamma \\ v\gamma & \gamma \end{bmatrix} \tag{4.24}$$

This is a linear transformation which relates one velocity to another velocity. The co-ordinate transformation is just a change of the ratio of the space axis to the time axis – a change of velocity[24]. These differently rotated co-ordinate systems are set over a 2-dimensional spinor space, the hyperbolic complex numbers, $\mathbb{S}$, and so special relativity is really spinor algebra mathematics. Hence the transformations in it are linear.

GR is set over an emergent expectation space. This is not a spinor space although all six of its 2-dimensional sub-spaces are spinor spaces. The co-ordinate transformations cannot be simple 4-dimensional rotations because there are no 4-dimensional rotations within a 4-dimensional emergent expectation space. Nor will the transformations be linear; if the transformations were linear, we would have a spinor algebra. However, we still have co-ordinate transformations over this type of space. Because GR is not set over a spinor space, in general, co-ordinate transformations within GR will not be linear but will be curvi-linear.

In a nutshell, special relativity deals with linear co-ordinate transformations in a spinor space and GR deals with non-linear co-ordinate transformations in an emergent expectation space. Riemann mathematics is the mathematics of an emergent expectation space.

---

[24] See: Dennis Morris *Empty Space is Amazing Stuff – The Special Theory of Relativity*.

Chapter 5

The Mathematics of Riemann Space II

We continue to develop the mathematics of Riemann space. "Once more into the breach...".

Tensors:

Once we have a definition of length, a distance function, a vector is the same length in all co-ordinate systems. Once we have a definition of 2-dimensional angle, the 2-dimensional angle between two vectors is the same in all co-ordinate systems. The components of a vector change from one co-ordinate system to another, but the vector, which represents something that really exists, does not change. Something which really exists will not change to suit humankind's arbitrary choice of co-ordinate system. A vector is an example of a tensor; it is a rank one tensor. In general, some properties of tensors, like a vector's length, are invariant (do not change) under change of co-ordinates even though their components do change under change of co-ordinates. The physics of the system is within these invariant properties. This means that the equations of classical physics ought to be written in tensors[25]. We endeavour to do this. This is why physicists are enamoured with tensors.

If a tensor equation is true in one co-ordinate system, it is true in all co-ordinate systems. Tensor equations express the invariant properties of the tensor; that is why they are true in all co-ordinate systems.

There are not many tensors in GR. We have the metric tensor and its inverse; we have the Ricci tensor; we have the Riemann curvature tensor, and we have the energy-momentum tensor. We also meet the Kronecker delta tensor and, occasionally, the reader might meet the Levi-Civita tensor (symbol).

---

[25] Tensors of rank 2 or higher do not exist in spinor algebras; hence the 'classical' adjective.

Aside: The Levi-Civita tensor (symbol) is:

$$\varepsilon_{abcd} = \begin{cases} +1 & \text{for even permutations of} \quad abcd \\ -1 & \text{for odd permutations of} \quad abcd \\ 0 & \text{otherwise} \end{cases} \qquad (5.1)$$

The metric tensor, the inverse metric tensor, the Kronecker delta, and the Levi-Civita tensor (symbol) all transform as tensors but their components are unchanged in any inertial co-ordinate system in flat space-time. They are the only tensors with this property. The Levi-Civita tensor, for technical reasons is not actually a tensor, and so we sometimes call it the Levi-Civita symbol. The Levi-Civita symbol and the Levi-Civita connection are named after Tullio Levi-Civita (1873-1941).

But what is a tensor?:

A tensor is a tensor product of vectors. The tensor product is denoted by the $\otimes$ symbol. The components of a tensor are products of the components of vectors. We illustrate with 4-dimensional vectors; we have:

$$\begin{bmatrix} a \\ b \\ c \\ d \end{bmatrix} \otimes \begin{bmatrix} w \\ x \\ y \\ z \end{bmatrix} = \begin{bmatrix} aw & ax & ay & az \\ bw & bx & by & bz \\ cw & cx & cy & cz \\ dw & dx & dy & dz \end{bmatrix} \qquad (5.2)$$

Notice that the tensor product[26] of two vectors is not commutative unless the tensor is symmetric across the leading diagonal. We have:

$$\begin{bmatrix} w \\ x \\ y \\ z \end{bmatrix} \otimes \begin{bmatrix} a \\ b \\ c \\ d \end{bmatrix} = \begin{bmatrix} aw & bw & cw & dw \\ ax & bx & cx & dx \\ ay & by & cy & dy \\ az & bz & cz & dz \end{bmatrix} \qquad (5.3)$$

---

[26] The tensor product is sometimes called the outer product of two tensors.

We have presented the product of two vectors as a box of components. Following this direction, if we were to multiply three vectors together, we would have to present the product as a cube of components. In fact, squares or cubes of components are nothing to do with tensors, but they are a useful way to understand what is happening. A tensor, like a vector, is just a set of n-dimensionally ordered components (real numbers). We see that a vector is a tensor; so is a scalar (single number).

We say that a scalar is a rank zero tensor and that a vector is a rank one tensor. In general, the rank of a tensor is the number of vectors multiplied together to form it.

We can multiply two contravariant vectors together, in this case we form a contravariant tensor of rank two which we denote by $V^m \otimes V^n = T^{mn}$. Each mn refers to a particular component of the tensor. We can multiply two covariant vectors together, in this case we form a covariant tensor of rank two which we denote by $V_m \otimes V_n = T_{mn}$. We can multiply one contravariant vector and one covariant vector together, in this case we form a mixed tensor of rank two which we denote by $V^m \otimes V_n = T^m_n$.

Symmetric and anti-symmetric tensors:
Looking at (5.2) & (5.3), we see that, if:

$$ax = bw, \quad ay = cw, \quad az = dw \quad\quad (5.4)$$

then the tensors would be symmetrical about the leading diagonal. More importantly, the product of the two vectors would be the same in either order and the multiplication would be commutative. We call a tensor which is symmetric in its components a symmetric tensor. We write this as, notice the round brackets:

$$T^{(mn)} = T^{(nm)} \quad\quad (5.5)$$

Notice the positional order of the indices in (5.5). In general, any tensor can be used to form a symmetric tensor as:

$$H_{(mn)} = \frac{1}{2}\left(H_{mn} + H_{nm}\right) \qu\quad (5.6)$$

Similarly, in (5.2) & (5.3), if:

$$ax = -bw, \quad ay = -cw, \quad az = -dw$$
$$aw = bx = cy = dz = 0$$

(5.7)

The tensors (5.2) & (5.3) would be anti-symmetric across the leading diagonal. We write this as, notice the square brackets:

$$T^{[mn]} = -T^{[nm]}$$

(5.8)

In general, any tensor can be used to form an anti-symmetric tensor as:

$$H_{[mn]} = \frac{1}{2}\left(H_{mn} - H_{nm}\right)$$

(5.9)

We have:

$$H_{mn} = \frac{1}{2}\left(H_{mn} + H_{nm}\right) + \frac{1}{2}\left(H_{mn} - H_{nm}\right)$$

(5.10)

and so any tensor can be split into its symmetric and anti-symmetric parts. In due course, we will split the emergent expectation field tensor that emerges from the super-imposition of the A_3 fields into its symmetric part and its anti-symmetric part to form the energy-momentum tensor and the electromagnetic tensor.

The symmetric or anti-symmetric nature of a tensor is invariant under change of co-ordinate system; of course it is; the essence of a tensor is that its properties are the same in all co-ordinate systems.

Tensors as multi-linear mappings:
Rather frighteningly, tensors are multi-linear mappings from contravariant vectors and covariant vectors to the real numbers. All this really means is that we can extract real numbers from tensors by using the inner product. For example, let us map a contravariant vector and a covariant vector to a real number:

$$\begin{bmatrix} a & b & c & d \end{bmatrix} \begin{bmatrix} 1 & 0 & 0 & 0 \\ 0 & -1 & 0 & 0 \\ 0 & 0 & -1 & 0 \\ 0 & 0 & 0 & -1 \end{bmatrix} \begin{bmatrix} w \\ x \\ y \\ z \end{bmatrix} = aw - bx - cy - dz \quad (5.11)$$

This, (5.11), is just the space-time inner product of two vectors. The format shown is commonly used. In this case, we are in Cartesian co-ordinates[27] in flat space.

If the two vectors in (5.11) were the same vector, $\begin{bmatrix} t & x & y & z \end{bmatrix}$, we would have the space-time length of that vector given by $d^2 = t^2 - x^2 - y^2 - z^2$. In a different co-ordinate system, the square matrix in the middle would be different and the components of the vectors would be different, but the product of them as in (5.11) would produce the same real number for the length of the vector because length of a vector is invariant under change of co-ordinates. In a different type of space which measured length differently, say $d^2 = t^2 + x^2 + y^2 + z^2$, we would have a different square matrix in the middle and we would have a different real number produced by the mapping; this would be a different length; the different type of space would measure length differently. The square matrix in the middle is an expression of the distance function of the type of space in a particular co-ordinate system. It is called the metric tensor.

The metric tensor appears differently in different co-ordinate systems within the same space. The metric tensor differs from one type of space to another.

Again:
Let's do that last bit again. Tensors are mathematical objects that take a number of contravariant vectors and a number of covariant vectors (1-forms) and combine them together to produce a real number.

---

[27] Cartesian co-ordinates are sometimes called rectangular co-ordinates.

In general, we show this with 2-dimensional vectors. A tensor with components:

$$\begin{bmatrix} \alpha_1\beta_1 & \alpha_1\beta_2 \\ \alpha_2\beta_1 & \alpha_1\beta_2 \end{bmatrix} \qquad (5.12)$$

will take two vectors:

$$\begin{bmatrix} a \\ b \end{bmatrix}, \begin{bmatrix} c \\ d \end{bmatrix} \qquad (5.13)$$

and produce the real number:

$$\alpha_1\beta_1 ac + \alpha_1\beta_2 ad + \alpha_2\beta_1 bc + \alpha_2\beta_2 bd \qquad (5.14)$$

If $\alpha_1\beta_1 = \alpha_2\beta_2 = 1$ & $\alpha_1\beta_2 = \alpha_2\beta_1 = 0$, this is the dot-product of the two vectors. If $\alpha_1\beta_1 = \alpha_2\beta_2 = 0$ & $\alpha_1\beta_2 = -1$ & $\alpha_2\beta_1 = 1$, this is the cross product of the two vectors.

You can now forget all about multi-linear mappings except for frightening sociology students. You cannot forget about the metric tensor and how it produces the inner products of two vectors.

Hang on! We have above, (5.11), multiplied the metric tensor by two vectors exactly as if it were a matrix, but we are calling it a tensor. What kind of object is the metric tensor? The metric tensor is a matrix which transforms under co-ordinate transformations in the same way that a tensor transforms (see below). Anything, matrices, vectors, ice cream cones, which transform as a tensor transforms is called a tensor. And so, the metric tensor is really a matrix and it is a tensor as well because it transforms as a tensor transforms.

Co-ordinate transformation of tensors:
The transformation properties of a tensor are the product of the transformation properties of the vectors which are multiplied together to form that tensor. We have:

$$\left(T^{mn}\right)' = \frac{\partial y^m}{\partial x^p} V^p \frac{\partial y^n}{\partial x^q} W^q = \frac{\partial y^m}{\partial x^p} \frac{\partial y^n}{\partial x^q} V^p W^q$$

$$= \frac{\partial y^m}{\partial x^p} \frac{\partial y^n}{\partial x^q} T^{pq}$$

(5.15)

Within this expression, (5.15), the $\{p, q\}$ indices are summed over as is the Einstein convention; because they are summed over, they are called dummy indices. We also have

$$\left(T_{mn}\right)' = \frac{\partial x^p}{\partial y^m} V_p \frac{\partial x^q}{\partial y^n} W_q$$

$$= \frac{\partial x^p}{\partial y^m} \frac{\partial x^q}{\partial y^n} T_{pq}$$

(5.16)

and:

$$\left(T^n_m\right)' = \frac{\partial x^p}{\partial y^m} V_p \frac{\partial y^n}{\partial x^q} W^q$$

$$= \frac{\partial x^p}{\partial y^m} \frac{\partial y^n}{\partial x^q} T^q_p$$

(5.17)

Scalars transform with zero differentials. We have:

$$S' = S$$

(5.18)

This is simply saying that a single number is the same in all co-ordinate systems.

The Clifford product of vectors:
The Clifford product of two vectors is nothing to do with tensors, but presenting it will help the reader to fit tensors into the overall scheme of things.

The tensor product of two vectors is not the only way to multiply vectors together. There is the Clifford product of vectors. The Clifford product is a *bona fide* non-commutative spinor algebra multiplication, and we find it within the spinor algebras that derive from the $C_2 \times C_2 \times ...$ types of finite groups. These algebras contain double cover spinor rotations and are closely connected to quantum field theory, QFT.

In essence, Clifford algebra multiplication of vectors includes basis vectors which it takes to be square roots of either minus unity or plus unity, $\left(\vec{e_i}\right)^2 = \pm 1$ and it takes the products of basis vectors to be anti-commutative, $\overrightarrow{e_i e_j} = -\overrightarrow{e_j e_i} : i \neq j$ but also to be the square roots of plus or minus unity. We might have written (5.2) as:

$$\begin{bmatrix} \vec{ae_1} \\ \vec{be_2} \\ \vec{ce_3} \\ \vec{de_4} \end{bmatrix} \otimes \begin{bmatrix} \vec{we_1} \\ \vec{xe_2} \\ \vec{ye_3} \\ \vec{ze_4} \end{bmatrix} = \begin{bmatrix} \overrightarrow{awe_1 e_1} & \overrightarrow{axe_1 e_2} & \overrightarrow{aye_1 e_3} & \overrightarrow{aze_1 e_4} \\ \overrightarrow{bwe_2 e_1} & \overrightarrow{bxe_2 e_2} & \overrightarrow{bye_2 e_3} & \overrightarrow{bze_2 e_4} \\ \overrightarrow{cwe_3 e_1} & \overrightarrow{cxe_3 e_2} & \overrightarrow{cye_3 e_3} & \overrightarrow{cze_3 e_4} \\ \overrightarrow{dwe_4 e_1} & \overrightarrow{dxe_4 e_2} & \overrightarrow{dye_4 e_3} & \overrightarrow{dze_4 e_4} \end{bmatrix} \quad (5.19)$$

If we were doing Clifford algebra, say the 16-dimensional $Cl_{4,0}$ in which the four basis vectors are square roots of plus unity and the products of two basis vectors are all square roots of minus unity[28], this product, (5.19), would be:

$$\begin{bmatrix} \vec{ae_1} \\ \vec{be_2} \\ \vec{ce_3} \\ \vec{de_4} \end{bmatrix} \begin{bmatrix} \vec{we_1} \\ \vec{xe_2} \\ \vec{ye_3} \\ \vec{ze_4} \end{bmatrix} = \begin{bmatrix} aw & -\sqrt{-1}ax & -\sqrt{-1}ay & -\sqrt{-1}az \\ \sqrt{-1}bw & bx & -\sqrt{-1}by & -\sqrt{-1}bz \\ \sqrt{-1}cw & \sqrt{-1}cx & cy & -\sqrt{-1}cz \\ \sqrt{-1}dw & \sqrt{-1}dx & \sqrt{-1}dy & dz \end{bmatrix}$$

$$(5.20)$$

Clearly, the Clifford product of two vectors is a different thing from the tensor product of two vectors. The Clifford product is not used in Riemann spaces because Riemann spaces have only real axes, $\mathbb{R}^n$, whereas the spaces of the Clifford algebras are spinor algebra spaces with one real axis and $(n-1)$ imaginary axes. The spinor algebra spaces are irrevocably flat.

---

[28] See: Dennis Morris The Naked Spinor – A Rewrite of Clifford Algebra.

We use the Clifford product of two vectors in spinor spaces like the 4-dimensional A_3 algebras, and we use the tensor product of two vectors in emergent expectation spaces like our 4-dimensional space-time.

The wedge product:
Although we do not use it in this book, the reader might meet the wedge product of covariant vectors (and of higher rank covariant tensors). We might think of this as the tensor commutator of two 1-forms (covariant vectors). Using the tensor product shown above, (5.3), the wedge product of two –forms is:

$$\alpha \wedge \beta = \alpha \otimes \beta - \beta \otimes \alpha \qquad (5.21)$$

We see:

$$\alpha \wedge \beta = -\beta \wedge \alpha$$
$$\alpha \wedge \alpha = 0 \qquad (5.22)$$

Chapter 6

The Metric Tensor

The metric tensor is a thing of central importance within Riemann geometry. From the metric tensor, with the Levi-Civita connection, we can calculate everything there is to know about a space except the global topology. It is because the metric tensor is local to each point in the space (different at each point) that we cannot use the metric tensor to calculate the global topological properties of the space.

The metric tensor encodes the geometry of space-time by expressing deviations from the Pythagorean theorem (in Cartesian co-ordinates). We measure curvature by how much the Pythagoras theorem is altered, and we call the change in the Pythagoras theorem curvature.

Above, we have, (5.19), the product of two vectors expressed as a box of components with basis vectors displayed. Within Clifford algebra, the products of the basis vectors are clearly defined as above. We might expect a similar definition of the products of the basis vectors within tensor calculus; we have one. The products of basis vectors are the inner products (dot products) of those vectors. In our 4-dimensional space-time these inner products are the inner products of basis 4-vectors. As we said above, the 4-vector inner product is fabricated out of the 2-dimensional inner products within the complex numbers, $\mathbb{C}$, and the hyperbolic complex numbers, $\mathbb{S}$.

We have:

$$
\begin{bmatrix} a\vec{e_1} \\ b\vec{e_2} \\ c\vec{e_3} \\ d\vec{e_4} \end{bmatrix} \otimes \begin{bmatrix} w\vec{e_1} \\ x\vec{e_2} \\ y\vec{e_3} \\ z\vec{e_4} \end{bmatrix} = \begin{bmatrix} aw(\vec{e_1}\cdot\vec{e_1}) & ax(\vec{e_1}\cdot\vec{e_2}) & ay(\vec{e_1}\cdot\vec{e_3}) & az(\vec{e_1}\cdot\vec{e_4}) \\ bw(\vec{e_2}\cdot\vec{e_1}) & bx(\vec{e_2}\cdot\vec{e_2}) & by(\vec{e_2}\cdot\vec{e_3}) & bz(\vec{e_2}\cdot\vec{e_4}) \\ cw(\vec{e_3}\cdot\vec{e_1}) & cx(\vec{e_3}\cdot\vec{e_2}) & cy(\vec{e_3}\cdot\vec{e_3}) & cz(\vec{e_3}\cdot\vec{e_4}) \\ dw(\vec{e_4}\cdot\vec{e_1}) & dx(\vec{e_4}\cdot\vec{e_2}) & dy(\vec{e_4}\cdot\vec{e_3}) & dz(\vec{e_4}\cdot\vec{e_4}) \end{bmatrix}
$$

$$(6.1)$$

There is a subtle assumption here. When we take the inner product of two basis vectors, we are assuming that the point at which we take these inner products is flat. We are assuming local flatness. The basis vectors are in the six 2-dimensional tangent planes[29]. We are also assuming the presence within the space of the 2-dimensional angles which correspond to the 2-dimensional inner products from which the 4-dimensional inner product is fabricated.

The reader might think that the inner product of two different basis vectors is zero; the reader would be incorrect to think this. If two Euclidean basis vectors, or Euclidean vectors in general, are orthogonal, then their inner product is zero.[30] We are not necessarily working in a co-ordinate system in which the axes are at Euclidean right-angles to each other, and so the inner product of two different basis vectors is not necessarily zero.

We want an inner product of two vectors within our Riemann space. We use (6.1) to give us one. We take (remember, this is not matrix multiplication):

$$\begin{aligned} V \cdot W &= V^m \vec{e}_m \cdot W^n \vec{e}_n \\ &= V^m W^n \left(\vec{e}_m \cdot \vec{e}_n \right) \\ &= V^m W^n g_{mn}(x) \end{aligned} \qquad (6.2)$$

Within this expression, we have:

$$g_{mn}(x) = \begin{bmatrix} \left(\vec{e}_1 \cdot \vec{e}_1\right) & \left(\vec{e}_1 \cdot \vec{e}_2\right) & \left(\vec{e}_1 \cdot \vec{e}_3\right) & \left(\vec{e}_1 \cdot \vec{e}_4\right) \\ \left(\vec{e}_2 \cdot \vec{e}_1\right) & \left(\vec{e}_2 \cdot \vec{e}_2\right) & \left(\vec{e}_2 \cdot \vec{e}_3\right) & \left(\vec{e}_2 \cdot \vec{e}_4\right) \\ \left(\vec{e}_3 \cdot \vec{e}_1\right) & \left(\vec{e}_3 \cdot \vec{e}_2\right) & \left(\vec{e}_3 \cdot \vec{e}_3\right) & \left(\vec{e}_3 \cdot \vec{e}_4\right) \\ \left(\vec{e}_4 \cdot \vec{e}_1\right) & \left(\vec{e}_4 \cdot \vec{e}_2\right) & \left(\vec{e}_4 \cdot \vec{e}_3\right) & \left(\vec{e}_4 \cdot \vec{e}_4\right) \end{bmatrix} \qquad (6.3)$$

We call $g_{mn}(x)$ the metric tensor. It is locally defined in the tangent space at each point in the emergent expectation space. In our 4-

---

[29] There are six 2-dimensional planes, three of each type of 2-dimensional space, in our 4-dimensional space-time.

[30] The inner product of two vectors in 2-dimensional space-time is never zero because the $\cosh(\ )$ function is never zero.

dimensional space-time, the tangent space is a fabrication of 2-dimensional spinor spaces. Note that, since the metric tensor is formed from the inner products of two basis vectors, it is a covariant tensor of rank two. The metric tensor is a matrix which transforms as a 2nd rank covariant tensor, and so it is a tensor and it is a matrix.

Notice that the metric tensor is a symmetric tensor because of the symmetry of the inner product, $\vec{e_i} \cdot \vec{e_j} = \vec{e_j} \cdot \vec{e_i}$, and so it has only ten independent components (in 4-dimensional space). In general, the metric tensor has $\dfrac{n(n+1)}{2}$ independent components in a space of n dimensions. The symmetry of the metric tensor is expressed as:

$$g_{\mu\nu} = g_{\nu\mu} \qquad (6.4)$$

There is a theorem of matrix algebra, well-known to people who know it well, that a co-ordinate transformation can always be found that will make a symmetric real matrix into a diagonal matrix with each entry on the main diagonal being either $\{+1, -1, 0\}$. This theorem essentially says that, since the metric tensor is a real symmetric matrix, the space is a flat space at any single point. At different points in the space, different co-ordinate transformations might be needed, and so, for a given co-ordinate transformation, the space is not necessarily flat everywhere.

In fact, at any point, p, of a Riemann space there is a co-ordinate system in which the metric tensor, $g_{\mu\nu}$, will be diagonal and the first derivatives all vanish. In our 4-dimensional space-time, such a co-ordinate system is a local inertial reference frame at the point p only. The second derivatives do not vanish.

Aside: Spectral theorem: For every symmetric real matrix, A, there exists an orthogonal matrix, O, such that $O^T A O = D$ where D is a diagonal matrix.

For every pair of symmetric real matrices, $A \& B$, if $AB = BA$, the matrices can both be diagonal in the same co-ordinate system.

For every pair of symmetric real matrices, $A \& B$, if $AB = BA$, the product of the matrices is symmetrical.

For every symmetric real matrix, A, A'' is symmetrical if and only if $\det(A) \neq 0$.

Because the metric tensor is symmetric, to transform it into its diagonal form in 4-dimensional space requires us to find only ten variables (components) in the 4×4 transformation matrix, but, because there are sixteen components in a 4×4 matrix, we have sixteen equations with which to find these ten variables. This leaves us six degrees of freedom; all of which preserve the metric tensor (vector length). A degree of freedom which preserves a length is a rotation. We have six rotations in our 4-dimensional space-time, three Lorentz boosts and three Euclidean rotations. See how well it fits together.

If we use Cartesian co-ordinates and are in flat Euclidean space, then the inner product of two different basis vectors is zero and the inner product of a basis vector with itself is unity. In that case, the metric tensor becomes:

$$g_{mn}(x) = \begin{bmatrix} 1 & 0 & 0 & 0 \\ 0 & 1 & 0 & 0 \\ 0 & 0 & 1 & 0 \\ 0 & 0 & 0 & 1 \end{bmatrix} \tag{6.5}$$

This corresponds to a distance function of the quadratic form:

$$ds^2 = dw^2 + dx^2 + dy^2 + dz^2 \tag{6.6}$$

Note that the components of the metric tensor can vary from point to point within the space as indicated by the (x) in $g_{mn}(x)$. This is equivalent to variation from point to point of the co-ordinate system in which the metric tensor is diagonal.

The components of the metric tensor will vary from point to point in a space because the properties of the co-ordinates system, or the properties of the space, or the properties of both, vary from point to point in the space. The metric tensor depends upon both the co-ordinate system used and upon the curvature of the underlying space.

Some examples of the metric tensor:

Let us consider a co-ordinate transformation from Cartesian co-ordinates to oblique rectilinear co-ordinates. In Cartesian co-ordinates, the quadratic form (6.6), has no mixed terms like $dwdx$. In oblique rectilinear co-ordinates, the quadratic form becomes a general quadratic form with such mixed terms. The coefficients of the elements in the quadratic form is just the metric tensor:

$$dw^2 + dx^2 + dy^2 + dz^2 \rightarrow dw^2 + dwdx + dwdy + ... + dydz + dz^2$$
$$= g_{\mu\nu} dx^\mu dx^\nu$$

$$(6.7)$$

The length given by (6.6) is invariant under change of co-ordinates. It follows that $g_{\mu\nu} dx^\mu dx^\nu$ is that same invariant. Thus, for a vector, $\vec{A}$, the length of that vector given by $g_{\mu\nu} A^\mu A^\nu$ is invariant. We can, and many do, think of the metric tensor as a generalised form of the inner product.

A Riemann metric (quadratic form) always transforms into a Riemann metric provided the co-ordinate transformation is differentiable and non-singular.

In 2-dimensional polar co-ordinates, the metric tensor is (4.12):

$$\begin{bmatrix} \vec{e}_r \cdot \vec{e}_r & \vec{e}_r \cdot \vec{e}_\theta \\ \vec{e}_\theta \cdot \vec{e}_r & \vec{e}_\theta \cdot \vec{e}_\theta \end{bmatrix} = \begin{bmatrix} 1 & 0 \\ 0 & r^2 \end{bmatrix} \qquad (6.8)$$

This corresponds to a distance function within the flat 2-dimensional space:

$$ds^2 = dr^2 + r^2 d\theta^2 + 0.drd\theta + 0.d\theta dr \qquad (6.9)$$

The metric tensor of the surface of a sphere is:

$$\begin{bmatrix} R^2 & 0 \\ 0 & R^2 \sin^2 \theta \end{bmatrix} \qquad (6.10)$$

This corresponds to the distance function within the spherical surface:

$$ds^2_{Sphere} = R^2 d\theta^2 + 0.d\theta d\phi + 0.d\phi d\theta + R^2 \sin^2\theta d\phi^2 \quad (6.11)$$

Notice that the components of the metric tensor correspond to the coefficients of the distance function as:

$$\begin{bmatrix} d\theta d\theta & d\theta d\phi \\ d\phi d\theta & d\phi d\phi \end{bmatrix} \quad (6.12)$$

This is entirely general.

The metric tensor of spherical co-ordinates is:

$$\begin{bmatrix} 1 & 0 & 0 \\ 0 & r^2 & 0 \\ 0 & 0 & r^2 \sin^2\theta \end{bmatrix} \quad (6.13)$$

The metric tensor of 4-dimensional flat Minkowski space-time is traditionally given a symbol of its own rather than $g_{\mu\nu}$. We have:

$$\eta_{\mu\nu} = \begin{bmatrix} 1 & 0 & 0 & 0 \\ 0 & -1 & 0 & 0 \\ 0 & 0 & -1 & 0 \\ 0 & 0 & 0 & -1 \end{bmatrix} \quad (6.14)$$

Note that, although the 2-dimensional flat space of special relativity is a spinor space, the 4-dimensional flat Minkowski space-time is an emergent expectation space. These are two very different types of space.

Yet more about the metric tensor:
We can think of the metric tensor as a product of 1-forms (covariant vectors) which accepts two contravariant vectors as arguments and spits out the inner product of those contravariant vectors. This is why the components of the tensor are defined as $\vec{e_i} \cdot \vec{e_j}$.

Another way to think of the metric tensor is as a means of transforming from one co-ordinate system to another and back again. That is transforming contravariant vectors into covariant vectors and covariant vectors into contravariant vectors. The metric tensor adjusts from contravariant transformation to covariant transformation – it reverses, or un-reverses, the direction of the co-ordinate change.

It is often said that, when all is said and done, in the view of Riemann geometry, the metric tensor is $g_{\mu\nu} = \vec{e}_\mu \cdot \vec{e}_\nu$ by human definition. Your author opines that, although it might be human observation which extracts the emergent expectation space from the A_3 spinor spaces, the metric tensor arises naturally from the presence of the 2-dimensional inner products in this emergent expectation space. The 'definition' of the metric tensor makes sense in an emergent expectation space with a quadratic expectation distance function because such a space accepts the 2-dimensional angles and associated inner products.

The determinant of the metric tensor:
Since the metric tensor is a matrix, it has a determinant. The determinant of the metric tensor is usually denoted by:

$$g = \det\left(g_{\mu\nu}\right) \qquad (6.15)$$

This determinant can never be zero because, if it were, then the four axes would not provide independent directions in space-time and would not be suitable as axes. In Cartesian co-ordinates, in our 4-dimensional space-time, the diagonal elements of the metric tensor are $\{1, -1, -1, -1\}$; all other elements are zero, and the determinant is thus $g = -1$. For any continuous change of co-ordinate system, like from Cartesian co-ordinates to oblique rectilinear co-ordinates, the determinant will change, but it will never pass through zero and will always be negative. We give an example of a 2-dimensional 'metric tensor' with zero determinant:

$$\begin{bmatrix} 1 & 2 \\ 2 & 4 \end{bmatrix} \sim \begin{bmatrix} dx^2 & dxdy \\ dydx & dy^2 \end{bmatrix} \qquad (6.16)$$

This is the metric of the distance function:

$$(dx + 2dy)^2 = d^2 \qquad (6.17)$$

Which is really a 1-dimensional space rather than a 2-dimensional space.

But where does the inner product come from?:
The standard answer within Riemann geometry to the question "Where does the inner product come from?" is "Out of thin air. We just invent it". For example; the inner product of 4-dimensional space-time is:

$$\begin{bmatrix} a \\ b \\ c \\ d \end{bmatrix} \cdot \begin{bmatrix} w \\ x \\ y \\ z \end{bmatrix} = aw - bx - cy - dz \qquad (6.18)$$

The actual origin of the inner product is within the two 2-dimensional spinor algebras the complex numbers, $\mathbb{C}$, and the hyperbolic complex numbers, $\mathbb{S}$. Only spinor algebras possess true inner products. (We simply multiply a complex number by its conjugate to produce a real number.) The 4-dimensional 4-vector inner product is a fabrication of six 2-dimensional inner products.

What exactly is the metric tensor?:
The metric tensor is a measure of the 2-dimensional angles between vectors (inner products) and the lengths of vectors (inner products). By adjusting the components of the metric from point to point over a stretched, distorted and curved Riemann space, we can stretch or squash the space in such a way as to ensure that the length of a vector is the same at all points in the space and that the 2-dimensional angles between two vectors are the same at all points in space. Oh! if only we had a vector or two of set length and of set angle between

them; we do; such vectors are found in only spinor algebra spaces; we have such vectors and angles in the 2-dimensional spinor algebra spaces which fit into our space-time as sub-spaces.

The length, as measured by the appropriate distance function, of a vector is a real number, a scalar, a rank zero tensor. A scalar has the same value in all co-ordinate systems. Thus, if the vector represents something which really exists, the length of a vector should be the same in all co-ordinate systems. The 2-dimensional angle between two vectors is also a scalar; this too should be the same in all co-ordinate systems. How do we arrange for the length of a vector and the angle between two vectors to be invariant within a space that might be stretched and distorted? We stretch and distort the space to fit our requirements and call it curvature.

Transformation of the metric tensor:

Transforming the basis 1-forms dx^i gives:

$$dx^m = \frac{\partial x^m}{\partial y^p} dy^p \quad \& \quad dx^n = \frac{\partial x^n}{\partial y^q} dy^q \qquad (6.19)$$

This leads to:

$$g'_{pq} = \frac{\partial x^m}{\partial y^p} \frac{\partial x^n}{\partial y^q} g_{mn} \qquad (6.20)$$

and we see that the metric tensor transforms as a covariant tensor of rank two. Because the metric tensor transforms as a covariant tensor of rank two, it is a covariant tensor of rank two. Tensors are defined by their transformation properties.

A matrix field:

The reader might sometimes hear the metric tensor, g_{ij}, referred to as a matrix field over a manifold, and the metric tensor is a matrix. A matrix field, g, can be a metric tensor if:

a) All 2nd derivatives of g exist and are continuous
b) g is not singular

c) g is symmetric

d) The distance function given by the diagonal elements of g is invariant under change of co-ordinate system.

The peculiarity of Riemann distance functions:

Within Riemann type spaces, we have distance functions which are quadratic forms like $d^2 = x^2 + y^2 + \ldots$. Such spaces can be covered by a variety of co-ordinate systems; examples are polar co-ordinates, cylindrical co-ordinates or spherical co-ordinates. We can bend the rules a little and allow a minus sign or two to creep into the distance function, but we do not allow distance functions that are not quadratic forms. An example of a non-quadratic emergent expectation distance function is the distance function which emerges from the two isomorphic 3-dimensional C_3 spinor algebras, $d^3 = a^3 + 3abc$.

Why do mathematicians restrict themselves to Riemann distance functions? History and observation of our 4-dimensional space-time. The 2-dimensional spaces have Riemann type distance functions; our space-time has a Riemann type of distance function; the three spatial dimensions of our space-time have a Riemann distance function. It is unwittingly assumed that all spaces of dimension higher than four should have a Riemann type of distance function because it is, unjustifiably, felt that our space-time should be a sub-space of those higher dimensional spaces using any set of four axes we choose. It seems that reality is not like this. Of all the emergent expectation distance functions, it seems that only our 4-dimensional space-time and the 2-dimensional spaces have a Riemann type of distance function.

In short, a Riemann space is based on the blatant assertion that:

$$ds^2 = \sum_{mn} g_{mn}(x) dx^m dx^n \tag{6.21}$$

No attempt is made to justify this assertion of a quadratic distance function. To Riemann geometers, this works by 'luck' in our 4-dimensional space-time. We assert that it actually works because the A_3 emergent expectation distance function is of this form.

Chapter 7

The Mathematics of Riemann Space III

Tensor operations:
Tensors can be added or subtracted provided they are of the same type and same size. We simply add the components. We cannot add a covariant tensor to a contravariant tensor any more than we can add two vectors written in different bases.

Tensors can be multiplied together. We have demonstrated this above, (5.2) & (5.3). Of course, multiplication is not commutative, and the tensors have to be of the same size.

Contraction of tensors:
Contraction is taking an upper index, putting it equal to a lower index, and summing over that index. It is a generalisation of the inner product. It is sometimes called taking the inner product of two tensors. We have the basic tensor co-ordinate transformation:

$$\left(V^m W_n\right)' = \frac{\partial y^m}{\partial x^a} \frac{\partial x^b}{\partial y^n}\left(V^a W_b\right) \tag{7.1}$$

If we put $m = n$, we get:

$$\left(V^m W_n\right)' = \frac{\partial x^b}{\partial y^a}\left(V^a W_b\right) = \delta_a^b\left(V^a W_b\right) \tag{7.2}$$

$$= V^a W_a \in \mathbb{R}$$

$\dfrac{\partial x^b}{\partial y^a}$ is zero when $a \neq b$ and is 1 when $a = b$. δ_a^b is called the

Kronecker delta. If is effectively the identity and plays a role similar to the number 1 in the real numbers.

Although we have used two vectors above and have come to the inner product of those vectors, the operation of contraction applies to all tensors that have both upper and lower indices. It makes no sense to contract together a pair of upper indices or a pair of lower indices; we must have an upper and a lower index. We will see shortly that the positions of indices can be swapped by use of the metric tensor; after being swapped, they can be contracted if the swapping has produced an upper and a lower index.

The inverse metric tensor:
Since the metric tensor has a non-zero determinant, it has an inverse such that:

$$g_{mn}g^{mp} = \delta_n^p \qquad (7.3)$$

Where δ_n^p is the Kronecker delta mentioned above which is effectively the identity. We have:

$$\delta_{ij} = \delta_j^i = \delta^{ij} \qquad (7.4)$$

Looking at the above, (7.3), we see that the inverse metric tensor is a second rank contravariant tensor, g^{mp}.

Above, (6.13), we gave the metric of spherical co-ordinates. With a little mathematics, we can calculate the inverse metric in spherical co-ordinates:

$$g^{rr}g_{rr} = 1 \Rightarrow g^{rr} = 1$$

$$g^{\theta\theta}g_{\theta\theta} = 1 \Rightarrow g^{\theta\theta} = \frac{1}{r^2} \qquad (7.5)$$

$$g^{\phi\phi}g_{\phi\phi} = 1 \Rightarrow g^{\phi\phi} = \frac{1}{r^2 \sin^2 \theta}$$

Giving:

$$g^{\mu\nu} = \begin{bmatrix} 1 & 0 & 0 \\ 0 & \dfrac{1}{r^2} & 0 \\ 0 & 0 & \dfrac{1}{r^2 \sin^2 \theta} \end{bmatrix} \qquad (7.6)$$

Raising and lowering indices:

We use the metric tensor, or its inverse, to raise and lower indices – to change a contravariant vector into a covariant vector and vice-versa. We do this by contracting the vector with the metric tensor:

$$V^m g_{mn} = V_n \quad \& \quad V_n g^{mn} = V^m$$
$$T^{ijk} = g^{ir} T_r^{jk} \dots\dots \tag{7.7}$$

We note that $V_\mu = V^\mu$ in only Cartesian co-ordinates in flat space.

The metric tensor again:

It is often said that, when all is said and done, the relationship between covariant vectors and contravariant vectors, $V_\mu = g_{\mu\nu} V^\nu$, is by human definition. Thinking back to our initial discussion of covariant and contravariant vectors when we simply rotated the co-ordinates, (4.1) & (4.2), your author opines that the relationship between the metric tensor and changing the type of vectors is nothing to do with human definition but arises through swapping from one co-ordinate transformation matrix to the other. We see that it is intimately involved in the angles between vectors and therefore in the inner products of those vectors and therefore in the lengths of those vectors. It ought not to be surprising that an object (the metric tensor) whose components are the inner products of vectors is able to convert rotation in one direction into rotation in the opposite direction.

Differentiation of vectors:

To differentiate a vector at a point in space, we begin by assuming the space at the point is locally flat. This local flatness is inherited from the underlying spinor algebras. We need a flat space because differentiation is a linear operation. The technical term for assuming a space is locally flat is to say that we are assuming Gaussian co-ordinates at that point. Gaussian co-ordinates are not necessarily Cartesian co-ordinates, they can wiggle about wildly, but they are flat co-ordinates. Gaussian co-ordinates are sometimes called normal co-ordinates.

Because the Gaussian co-ordinates are locally flat, the metric tensor does not vary from point to point in the infinitesimally small point – well, I'm not sure that makes sense, but it is the standard mantra. Alternatively, we take the view that we are working in the flat tangent space at the point.

In some co-ordinate systems, the basis vectors vary in direction and size from point to point in the space. This is not necessarily a property of the space but might be a property of the co-ordinate system. An example is polar co-ordinates over the 2-dimensional Euclidean plane. We have the radial basis vector $\vec{e}_r$ and the angular basis vector $\vec{e}_\theta$. Following the convention, these basis vectors have subscripted indices, of course. We have the polar co-ordinates:

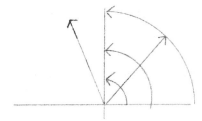

We see that both the length and the direction of the angular basis vector varies with distance from the origin and the direction of the radial basis vector varies with angle.

Because the basis vectors vary, differentiating only the components of a vector will not give the complete differential of the vector; we need to also differentiate the basis vectors. In fact, even if the basis vectors do not vary, we ought to differentiate them when we differentiate a vector. In elementary vector calculus, we do not differentiate a basis vector because differentiating a constant basis vector gives zero.

Differentiating a vector in Cartesian co-ordinates (with constant basis vectors) gives:

$$\frac{\partial}{\partial x}\left(f(x,y)\vec{e}_x + g(x,y)\vec{e}_y\right) = \frac{\partial f}{\partial x}\vec{e}_x + \frac{\partial \vec{e}_x}{\partial x}f + \frac{\partial g}{\partial x}\vec{e}_y + \frac{\partial \vec{e}_y}{\partial x}g$$

$$= \frac{\partial f}{\partial x}\vec{e}_x + 0 + \frac{\partial g}{\partial x}\vec{e}_y + 0$$

(7.8)

With a similar result for differentiation with respect to y. We see that, because the basis vectors do not vary, their differentials are zero, and so they are traditionally ignored.

Differentiating a vector in polar co-ordinates (with non-constant basis vectors) is different in that we have to account for the non-constant basis vectors. We have:

$$\frac{\partial \vec{e_r}}{\partial r} = \frac{\partial}{\partial r}\left(\cos\theta.\vec{e_x} + \sin\theta.\vec{e_y}\right)$$
$$= 0 \tag{7.9}$$

But:

$$\frac{\partial \vec{e_r}}{\partial \theta} = \frac{\partial}{\partial \theta}\left(\cos\theta.\vec{e_x} + \sin\theta.\vec{e_y}\right)$$
$$= -\sin\theta.\vec{e_x} + \cos\theta.\vec{e_y} \tag{7.10}$$
$$= \frac{1}{r}\vec{e_\theta}$$

Similarly:

$$\frac{\partial \vec{e_\theta}}{\partial r} = \frac{\partial}{\partial r}\left(-r\sin\theta.\vec{e_x} + r\cos\theta.\vec{e_y}\right)$$
$$= 0.\vec{e_r} + \frac{1}{r}\vec{e_\theta} = \frac{1}{r}\vec{e_\theta} \tag{7.11}$$

And:

$$\frac{\partial \vec{e_\theta}}{\partial \theta} = \frac{\partial}{\partial \theta}\left(-r\sin\theta.\vec{e_x} + r\cos\theta.\vec{e_y}\right)$$
$$= -r\vec{e_r} + 0.\vec{e_\theta} = -r\vec{e_r} \tag{7.12}$$

We notice that the differentials of the basis vectors are a multiple of a basis vector. In general, the differentials of the basis vectors are vectors and are therefore a sum of multiples of the different basis vectors; we have shown this by including the zeros above, (7.11) & (7.12).

In general, if we differentiate a vector, $\vec{V}$, with components, V^α, with respect to a co-ordinate, x^β, we will get a sum of the differentials of each co-ordinate multiplied by the appropriate basis vector, $\dfrac{\partial V^\alpha}{\partial x^\beta}\vec{e_\alpha}$, and the component of the vector multiplied by the differential of the appropriate basis vector, $V^\alpha \dfrac{\partial \vec{e_\alpha}}{\partial x^\beta}$. Note that we have used the summation convention in the previous sentence. The differential of a vector is:

$$\frac{\partial \vec{V}}{\partial x^\beta} = \frac{\partial V^\alpha}{\partial x^\beta}\vec{e_\alpha} + V^\alpha \frac{\partial \vec{e_\alpha}}{\partial x^\beta} \tag{7.13}$$

This is just the familiar product rule of differentiation.

Christoffel symbols:
We introduce a symbol for the differentials of the basis vectors:

$$\frac{\partial \vec{e_\alpha}}{\partial x^\beta} = \Gamma^\mu_{\alpha\beta}\,\vec{e_\mu} \tag{7.14}$$

Notice that there is summation over the μ. The $\Gamma^\mu_{\alpha\beta}$ is called a Christoffel symbol. It is named after Elwin Bruno Christoffel (1829-1900). A Christoffel symbol is the μth component of the vector that results from differentiating the α basis vector with respect to the β co-ordinate.

The Christoffel symbols, $\Gamma^\mu_{\alpha\beta}$, are a set of $n \times n$ matrices, $\Gamma^\mu_{\sim\beta}$ where n is the dimension of the space; there are n of them corresponding to the values of α. Thus, the Christoffel symbols are linear transformations. If the covariant derivative is going to be a linear operation and satisfy the Leibnitz rule for products, then it must differ from the partial derivative in only linear terms.

Within the 2-dimensional polar co-ordinates, the indices, $\{\alpha, \beta, \mu\}$ can take one of two values. Thus, there are $2^3 = 8$ different combinations of indices and so eight Christoffel symbols; because the metric is symmetric, only six of these are independent. In 2-dimensional polar co-ordinates, the Christoffel symbols are:

$$\Gamma^r_{rr} = 0 \quad \Gamma^r_{r\theta} = 0 \quad \Gamma^r_{\theta r} = 0 \quad \Gamma^r_{\theta\theta} = -r$$

$$\Gamma^\theta_{rr} = 0 \quad \Gamma^\theta_{r\theta} = \frac{1}{r} \quad \Gamma^\theta_{\theta r} = \frac{1}{r} \quad \Gamma^\theta_{\theta\theta} = 0$$

(7.15)

Note that a particular Christoffel symbol can be zero in some co-ordinate system but not zero in other co-ordinate systems, and that a flat space can have non-zero Christoffel symbols in some co-ordinate systems.

Clearly, in 4-dimensions, there are $4^3 = 64$ different combinations of indices and thus sixty-four Christoffel symbols; the symmetric form of the metric means that only forty of these are independent. Christoffel symbols are also known as connection coefficients or affine connection coefficients or an affine connection.

An affine connection on a space is usually thought of as defining what is meant by the parallel transport of a vector. In a curved space, the vector will generally change as it is transported from place to place but what it changes into will be another vector that is a sum of amounts of the basis vectors in the space. The Christoffel symbols are the amounts, the coefficients of the basis vectors, which measure the change during the transport of a vector. If the covariant differential (see below) of a vector is zero as it is transported along a path in the space, we say the vector has been parallel transported. Christoffel symbols, affine connection, parallel transport – same thing.

Within Riemann geometry, it is assumed that the Christoffel symbols are symmetric in their lower indices. This means:

$$\Gamma^\mu_{\alpha\beta} = \Gamma^\mu_{\beta\alpha}$$

(7.16)

Since the theory of GR is based in Riemann geometry, GR also makes this assumption.

Note that the Christoffel symbols are not tensors; I know they look like tensors, but they do not transform as tensors transform under a change of co-ordinate systems and so they are not tensors. Although the Christoffel symbols are not tensors, we can use each of them to construct a tensor called the torsion tensor:

$$Torsoin = T_{bc}^a = \Gamma_{bc}^a - \Gamma_{cb}^a \qquad (7.17)$$

Aside: There is an alternative theory of gravity produced by Ellie Cartan in which $\Gamma_{\alpha\beta}^\mu \neq \Gamma_{\beta\alpha}^\mu$. This imbalance of the Christoffel symbols is called torsion, and Cartan's theory of gravity is sometimes known as the torsion theory of gravity. We look at it in more detail later.

Roughly, the Christoffel symbols correspond to the gravitational force.

The covariant derivative:
Using the Christoffel symbols, the derivative of a contravariant vector is:

$$\frac{\partial \vec{V}}{\partial x^\beta} = \frac{\partial V^\alpha}{\partial x^\beta} \vec{e_\alpha} + V^\alpha \ \Gamma_{\alpha\beta}^\mu \ \vec{e_\mu} \qquad (7.18)$$

We are allowed to replace dummy indices with a different letter giving:

$$\frac{\partial \vec{V}}{\partial x^\beta} = \left(\frac{\partial V^\alpha}{\partial x^\beta} + V^\mu \ \Gamma_{\mu\beta}^\alpha \right) \vec{e_\alpha} \qquad (7.19)$$

The components of the vector field are then:

$$D_\beta (V^\alpha) = \frac{\partial V^\alpha}{\partial x^\beta} + V^\mu \ \Gamma_{\mu\beta}^\alpha \qquad (7.20)$$

This is called the covariant derivative. In some co-ordinate systems, this will reduce to the derivative of a vector which we learned in kindergarten, but it is the proper derivative of a contravariant vector in all co-ordinate systems. The word 'covariant' in this sense is not connected to covariant vectors.

Notice that the covariant derivative is not just something that occurs in only curved space. The above polar co-ordinates are over a flat space.

The covariant derivative of a covariant vector is given by:

$$D_\beta \left(V_\alpha \right) = \frac{\partial V_\alpha}{\partial x^\beta} - V_\mu \, \Gamma^\mu_{\alpha\beta} \qquad (7.21)$$

Note the sign difference. The sign before the Christoffel symbol is convention; we abide by the standard convention.

We denote the partial derivative, $\dfrac{\partial}{\partial x^\mu}$, as ∂_μ. The co-variant derivative with respect to x^μ is denoted in various ways, but we will denote it by D_μ.

Aside: The partial derivative of a vector is sometimes denoted by a comma, $\dfrac{\partial V^\alpha}{\partial x^\beta} = V^\alpha{}_{,\beta}$, and the covariant derivative is similarly denoted by a semi-colon $V^\alpha{}_{;\beta} = V^\alpha{}_{,\beta} + V^\mu \Gamma^\alpha_{\mu\beta}$. Thus covariant differentiation is sometimes known as semi-colon differentiation.

Aside: The covariant derivative of a vector is often denoted by $\overrightarrow{\nabla V}$.

We will be using the covariant derivative, rather than the partial derivative, when we write the field equations of GR. This is because, in general, the covariant derivative of a vector, or any other tensor, is itself a tensor and because the covariant derivative is the only proper derivative.

The covariant derivative of a scalar field:
The covariant derivative of a scalar field is the same as the partial derivative of a scalar field – the gradient. The scalar field does not depend upon the co-ordinate system, it is simply a value at each point, and so its differential does not vary with different basis vectors.

Differentiation of tensors:

We now know how to differentiate a contravariant vector and how to differentiate a covariant vector. A tensor is just a tensor product of vectors. Remembering that each $T_{\mu\nu}$ is just a single component of the covariant tensor, we have:

$$D_\beta T_{\mu\nu} = \frac{\partial T_{\mu\nu}}{\partial x^\beta} - T_{\alpha\nu}\Gamma^\alpha_{\mu\beta} - T_{\mu\alpha}\Gamma^\alpha_{\nu\beta}$$

$$D_\beta T^{\mu\nu} = \frac{\partial T^{\mu\nu}}{\partial x^\beta} + T^{\alpha\nu}\Gamma^\mu_{\alpha\beta} + T^{\mu\alpha}\Gamma^\nu_{\alpha\beta} \qquad (7.22)$$

$$D_\beta T^\mu_\nu = \frac{\partial T^\mu_\nu}{\partial x^\beta} + T^\alpha_\nu\Gamma^\mu_{\alpha\beta} - T^\mu_\alpha\Gamma^\alpha_{\nu\beta}$$

Higher rank tensors are differentiated similarly. We add one Christoffel term for each contravariant index and we subtract one Christoffel term for each covariant index.

Christoffel symbols again:

We will see in the next chapter that the covariant differential of the metric tensor is always zero. This leads to:

$$D_s g_{mn} = \frac{\partial g_{mn}}{\partial x^s} - \Gamma^t_{sm}g_{tn} - \Gamma^t_{sn}g_{tm} = 0 \quad ....(1)$$

$$D_m g_{sn} = \frac{\partial g_{sn}}{\partial x^m} - \Gamma^t_{sm}g_{tn} - \Gamma^t_{mn}g_{st} = 0 \quad ....(2) \qquad (7.23)$$

$$D_n g_{sm} = \frac{\partial g_{sm}}{\partial x^n} - \Gamma^t_{sn}g_{tm} - \Gamma^t_{mn}g_{st} = 0 \quad ....(3)$$

From this $(3)+(2)-(1)$ leads to:

$$\partial_n g_{sm} + \partial_m g_{sn} - \partial_s g_{mn} = 2\Gamma^t_{mn}g_{st} \qquad (7.24)$$

Inverting the metric gives:

$$\Gamma^t_{mn} = \frac{1}{2}g^{st}\left[\partial_n g_{sm} + \partial_m g_{sn} - \partial_s g_{mn}\right] \qquad (7.25)$$

This is how we usually calculate Christoffel symbols.

The importance of (7.25) is not so much the calculation usage but the fact that the Christoffel symbols all derive from the metric tensor. Once we have the metric tensor, we have all the Christoffel symbols. This is an important point, and so we will say it again. Once we have the metric tensor, we have all the Christoffel symbols. In due course, we will see that once we have the metric tensor, we also have the Riemann curvature tensor; this also is an important point.

Even though the Christoffel symbols (connection coefficients) are not tensors, we can still raise or lower their indices with the metric tensor. We can multiply (7.25) by the metric tensor, effectively lowering an index, to get:

$$\Gamma^{t}_{mn} g_{st} = \frac{1}{2} g^{st} g_{st} \left[\partial_n g_{sm} + \partial_m g_{sn} - \partial_s g_{mn} \right]$$

$$\Gamma_{mns} = \frac{1}{2} g^{st} g_{st} \left[\partial_n g_{sm} + \partial_m g_{sn} - \partial_s g_{mn} \right]$$

(7.26)

In many places, such raising and lowering of the indices of the Christoffel symbols is considered to be bad behaviour.

The Christoffel symbol with all three indices lowered is sometimes referred to as a Christoffel symbol of the first kind while the Christoffel symbol with one upper index is sometimes referred to as a Christoffel symbol of the second kind.

Chapter 8

The Mathematics of Riemann Space IV

In this book, to this point, we have dealt with only flat spaces. However, the mathematics we have developed applies to any co-ordinate system. In particular, it applies to co-ordinate systems over curved spaces, but we need to acquaint the reader with the idea of empty space being curved.

The reader can immediately understand that the polar co-ordinate system over a flat space has a curved co-ordinate. A suitable curved space with a straight co-ordinate system would be exactly the same. The important point is that a curved co-ordinate system over a flat space will have the same mathematics as a straight co-ordinate system over a curved space. Thus, if we know how to use curved co-ordinates over flat space, we know how to work with curved space.

How do we know that a space is curved? Perhaps the apparent curvature is just our choice of co-ordinates. If, given a particular co-ordinate system, that system has to be changed from point to point in the space, then surely there is curvature in the underlying space. No, we might have picked the wrong co-ordinate system. All we know is that a flat co-ordinate system cannot be used globally over all space; this is what we call a curved space.

It is because the curved spaces we consider have local flatness that the mathematics of flat space co-ordinate systems apply to them. Without local flatness, the whole mathematical structure of Riemann geometry falls to bits.

Your author has no idea what a curved space without local flatness might look like and he opines that no such space exists, but it should be remembered that, although local flatness is intuitively obvious when we think of a curved space embedded in a higher dimensional flat space, intuition cannot with certainty be extrapolated into new realms. Within GR, and all Riemann geometry, conventionally local flatness is an assumption which no-one can justify other than to say that it seems to work.

Although it is often not explicitly stated, Riemann geometry assumes that the rotations within the space are 2-dimensional rotations and that the n-dimensional curvature within a space is fabricated from what are fundamentally 2-dimensional curvatures. Within the spinor spaces, for example quaternion space, we find higher dimensional rotations, and we might wonder why we have 2-dimensional rotations in a 4-dimensional Riemann space rather than the intuitively more obvious 4-dimensional rotation. Riemann geometry also assumes the inner products which form the metric tensor are fabricated from 2-dimensional inner products[31] which are associated with the $\cos(\ )$ function or the $\cosh(\ )$ function rather than with, say, a 3-dimensional or a 4-dimensional trigonometric function such as we find in the spinor spaces.

Intrinsic curvature:

There are two types of curvature in classical (non-spinor) geometry. They are called extrinsic curvature, which we mention below, and intrinsic curvature. Extrinsic curvature is not really curvature, and so, when we use the word curvature, we always mean intrinsic curvature unless we use the adjective extrinsic.

An example of intrinsic curvature is the curvature of a sphere. It is impossible to take a piece of a spherical surface and lay it flat upon a flat table without stretching and distorting the spherical surface. The curvature is intrinsic to the spherical surface. This contrasts with a rolled up piece of paper which can easily be made flat.

Intuitively, the surface of a sphere is flat at every infinitesimally small point of its surface. Mathematically, local flatness means that, at any infinitesimally small point in the space, a co-ordinate system can be found in which the metric tensor will be such that all its off-leading-diagonal components are zero but its leading diagonal components are non-zero. We say the metric is diagonal at that point in that co-ordinate system. It is important to realise that each point might require a different co-ordinate system for the metric to be diagonal at that point. This is what we call intrinsic curvature. Thinking of the spherical surface, we have to change the orientation

---

[31] To see a 3-dimensional inner product and a 4-dimensional inner product, see Dennis Morris: The Higher Dimensional Forms.

of the tangent plane (rotate the co-ordinate system) as we move from point to point on the sphere. We give three examples of a diagonal metric:

$$g_{\mu\nu} = \begin{bmatrix} 1 & 0 \\ 0 & 1 \end{bmatrix} \quad g_{\mu\nu} = \begin{bmatrix} 1 & 0 \\ 0 & r^2 \end{bmatrix} \quad g_{\mu\nu} = \begin{bmatrix} 1 & 0 & 0 & 0 \\ 0 & -1 & 0 & 0 \\ 0 & 0 & -1 & 0 \\ 0 & 0 & 0 & -1 \end{bmatrix} \quad (8.1)$$

If a space is globally flat then a single co-ordinate system can be found in which the metric tensor will be diagonal at every point of the space. The components of the metric tensor do not vary with position in a globally flat space; they are the same at all points in the space. The components of the metric tensor do vary with position in a non-flat space. This is mathematically what we mean by curved space.

If a space is globally curved then no co-ordinate system can be found in which the metric tensor will be diagonal at more than one point of the space.

Clearly, it is possible to invent a space which has flat portions and curved portions.

Within GR, gravitational tidal forces correspond to the curvature of space-time. Uniform gravitational fields (there are no uniform gravitational fields in the real universe) do not have gravitational tidal forces and are not associated with space-time curvature.

Theorema Egregium:
Possibly the greatest mathematician who ever lived was Johann Carl Friedrich Gauss (1777-1855). Apart from his achievements, he is all the more remarkable because, in a rigidly class divided society, he was born the son of a bricklayer and yet he became the leading mathematician of his day.[32] Gauss laid the foundations of differential geometry upon which Riemann constructed Riemann geometry. Circa 1825, Gauss proved a foundational mathematical theorem

---

[32] I presume bricklayers worked on price, and perhaps it was counting the bricks that led to Gauss's extraordinary ability.

which underpins all of Riemann geometry. The importance of this theorem is made evident by the name Gauss, possibly the king of mathematicians, gave the theorem; Gauss called the theorem *Theorma Egregium* – king of theorems.

Theorema Egregium states that the geometry of a curved space, the angles and distances within that space, can be measured within that space without reference to a higher dimensional embedding space. In other words, our 4-dimensional space-time can be curved without being embedded into a higher dimensional flat space. We picture the 2-dimensional curved surface of a sphere embedded within our space-time, and so we might think that any curved space has to be embedded within a higher dimensional space so that it can be curved within that higher dimensional space. Theorema Egregium says, "No, we do not need the higher dimensional space".

In fact, we now know of four separate errors in Gauss's proof of Theorema Egregium. The errors are all of a topological nature and have been corrected. The theorem still stands. However, it gives pause for thought that what we today consider to be mathematical proof might be sometime later found to be contain errors. Of course, the proof of Theorema Egregium is based on the same assumptions, like 2-dimensional angles etc., as Riemann geometry is based upon.

Extrinsic curvature:
If I take a flat 2-dimensional sheet of paper and I roll it into a cylinder, I have changed how the flat piece of paper is embedded in our 4-dimensional space-time. I can unroll the paper and it will be a flat sheet again. During the rolling and unrolling of the sheet of paper, never once did I have to stretch or distort the paper to form the cylinder or reform the flat sheet.

If I take a flat rubber sheet and I partially wrap it around a sphere, I have to stretch and distort the rubber sheet to make its shape coincide with the spherical surface.

The cylinder is curved only by virtue of the way it is embedded in our space-time. A single co-ordinate system can be found in which the 2-dimensional metric tensor of the cylindrical sheet is diagonal at every point on the sheet. We say the cylinder has extrinsic curvature but that it is intrinsically flat. No single co-ordinate system can be found in which the 2-dimensional metric tensor of the spherical surface is diagonal at more than one point on the surface. We say the sphere has intrinsic curvature; it is not intrinsically flat. The difference between extrinsic curvature and intrinsic curvature is whether or not a space can be flattened without distortion.

Extrinsic curvature is not really curvature. Intrinsic curvature is equivalent to saying that the metric tensor varies with position in the space. We are interested in only intrinsic curvature. When we speak of curved space, we will be speaking of intrinsically curved space. When we speak of curved space, we will be speaking of space in which the metric varies with position in the space.

A little history:

The view that our space-time is intrinsically curved rather than embedded in a flat 5-dimensional space is certainly the modern view, but it was not held by all noted physicists in the early 20th century. Paul Dirac (1902–1984), writing in his book 'General Theory of Relativity' published in 1975 writes:

"One can easily imagine a curved 2-dimensional space as a surface immersed in Euclidean 3-dimensional space. In the same way, one can have a curved 4-dimensional space immersed in a flat space of a larger number of dimensions. Such a curved space is called a Riemann space. A small region of it is approximately flat. Einstein assumed that physical space is of this nature and thereby laid the foundation for his theory of gravitation."

The idea of our 4-dimensional curved space-time being embedded in a higher dimensional flat space is not without some justification. Where does local flatness come from if we are not embedded in a higher dimensional flat space? The modern view chooses to assume local flatness rather than assume a flat embedding space. Unless we can find another source of local flatness, assuming local flatness is very much like assuming a higher dimensional flat embedding space.

Of course, your author opines that the local flatness of an emergent expectation space such as our 4-dimensional space-time is inherited from the underlying spinor algebras, and so your author does not have to assume a higher dimensional flat embedding space. Although the flatness is inherited at an infinitesimally small point, the curvature of a space is dependent upon the nature of the rotations in the space. In our 4-dimensional space-time, the rotations are 2-dimensional and so the curvature is a fabrication of 2-dimensional curvatures, but the local flatness is a 4-dimensional flatness. At an infinitesimally small point the nature of the curvature is irrelevant, and so we do not need 2-dimensional rotations to specify it.

Flatness and the metric tensor:

We cannot diagnose intrinsic curvature simply by looking at the metric tensor. It might be that the metric tensor is written in curved co-ordinates. In practice, we calculate the Riemann curvature tensor (see later) to determine whether or not a space is curved; a space is flat if the Riemann curvature tensor is zero; if the Riemann curvature tensor is not zero, the space is curved.

If a space is curved in one set of co-ordinates, it is equally curved in other sets of co-ordinates. This means that any measure of its curvature must be invariant under change of co-ordinates; any measure of curvature must be a tensor. This is why the curvature tensor is a tensor.

Riemann local flatness is 2-dimensional:

In conventional Riemann geometry, local flatness is a fabrication of 2-dimensional types of flatness and is associated with zero curvature. There are two types of flat 2-dimensional space; these are the Euclidean plane of the complex numbers, $\mathbb{C}$, and the flat 2-dimensional space-time of special relativity, the hyperbolic complex numbers, $\mathbb{S}$, and it is from these two types of flat spinor spaces that the conventional local flatness of Riemann geometry is fabricated. We differ from Riemann mathematics in that we associate local flatness with the underlying spinor spaces and consider it to be nothing to do with zero curvature.

Curvature in our space-time:

In our 4-dimensional space-time, we have curvature in a space with the distance function:

$$t^2 - x^2 - y^2 - z^2 \qquad (8.2)$$

This distance function is not the distance function of a spinor space and as such is not associated with a 4-dimensional rotation. This distance function cannot determine 4-dimensional rotation because this quadratic form is not preserved under multiplication and therefore cannot be the form of the determinant of a rotation matrix:

$$\left(t^2 - x^2 - y^2 - z^2\right)\left(a^2 - b^2 - c^2 - d^2\right) \neq K^2 - L^2 - M^2 - N^2 \quad (8.3)$$

It can, of course, allow either of the two types of 2-dimensional rotation from the two 2-dimensional spinor algebras.

In general, any distance function of the general quadratic form $Adx^2 \pm Bdxdy \pm Cdy^2 \pm Ddxdz \pm \ldots$ can be reduced by a co-ordinate transformation to a quadratic form with no mixed terms. The signature of a quadratic distance function (the numbers of plus and minus signs) will remain the same under any co-ordinate transformation – Sylvester's theorem.

It is a mathematical fact that if a matrix is symmetric, there is some co-ordinate transformation which will make that matrix diagonal:

$$\Lambda g_{mn} \Lambda = diag\left(\lambda_1, \lambda_2, \ldots\right) \qquad (8.4)$$

Since it is the symmetry of the 2-dimensional inner product that produces the symmetry of the metric tensor which thus produces local flatness, we might take the view that our space-time is locally flat because of the 2-dimensional spinor algebras.

Intrinsic curvature – another look:

Intrinsic curvature is variation of the components of the metric tensor, $g_{\mu\nu}$, from place to place in our space-time. Such variation is variation of the 4-dimensional fabrication of 2-dimensional inner products of two vectors that are the components of the metric tensor. Inner products, which are the length of a single vector or the angle between two vectors, are invariant scalars. If the inner products cannot vary, there must be something about the underlying space-time that is varying. Curvature is what we call this variation of the underlying space-time.

Alternatively, we could say that space-time is flat but that the inner products vary from point to point. This is to say that the length of a vector varies from place to place as does the angle between two vectors.

Draw an equilateral triangle on to a flat rubber sheet. The angles within the triangle are 60°. Now stretch the rubber sheet so that it fits around a spherical surface such that the base of the triangle lies along the equator of the sphere and the sides of the triangle lie along the Greenwich meridian and the 90° East longitude. The angle between the vectors has changed.

So, we do not need an intrinsically curved space-time, but, if we reject intrinsic curvature, we have to accept varying inner products.

The continuity of variation of angle between two vectors associated with local variation of the inner product implies local flatness, and local flatness implies continuity of local variation of inner product.

In the elementary calculus of finding maxima and minima, the first differential of a function can be zero but the second differentials give the curvature. Within Riemann geometry, even though the second differentials of the metric are not zero, because the first differentials are, we must have local flatness.

Straight lines in curved space:
The concept of a straight line is meaningless in curved space. Instead we have the concept of a geodesic. In flat space, a geodesic and a straight line are the same thing.

A geodesic is a route through space, a line if you like, which is either the shortest distance between two points or the longest distance between two points[33]. There might be many routes through a curved space which are equally the shortest, or longest, distance between two points. Thinking of a spherical surface, the Earth say, the lines between the south pole and the north pole are all geodesics of equal

---

[33] In the Lagrangian formulation, these are the stationary distances which can be maxima or minima.

shortest length. Similarly, any great circle around the Earth is the geodesic of longest length from a point to itself. There is not necessarily any unique shortest, or unique longest, distance between two points in a curved space. In the case of the geodesic being shortest possible distance between two points, we can think of a geodesic as being the route through the space along which a taut string would lie.

Ultimately, it is because the first differentials of the metric are zero that bodies move along geodesics.

Within Riemann spaces of all positive signature, this means the signs in the quadratic distance function are all pluses, a geodesic is the shortest possible route between two points. This is like a straight line in flat 2-dimensional Euclidean space being the shortest distance between two points.

Within our space-time, which has signature $(+,-,-,-)$, a geodesic is the longest possible line. This is like a straight line in flat 2-dimensional space-time being the longest distance between two points.[34]

Within GR, bodies move (fall if you prefer) along geodesics in 4-dimensional space-time. A planet orbiting the sun is in free-fall along a geodesic in space-time. As a body moves along a geodesic in a non-uniform gravitational field, it will experience tidal forces (see later). These tidal forces are associated with what we call gravity and are proportional to the Riemann curvature tensor.

Parallel transport of vectors:
Parallel transport of vectors is also called an affine connection.

In a flat space, we can move a vector around through any route we choose, and, when we return it to its starting point, the vector will still point in the same direction and be of the same length as it was before we moved it. In a curved space, this is not true. Consider a vector upon the surface of the Earth pointing due east on the Greenwich meridian at the equator. Let us move it around on the

---

[34] See: Dennis Morris *Empty Space is Amazing Stuff – The Special Theory of Relativity.*

surface of the Earth being careful not to change its direction. First, we move it up the Greenwich meridian to the north pole; it continues to point eastward until, at the north pole, it points southward. Second, keeping the vector's direction unchanged, move it southward down the 90° east longitude to the equator; it continues to point southward all the way. Third, move the vector back along the equator to Greenwich meridian. The vector continues to point southward, and we see that its direction has changed from eastward to southward by simply moving it around the surface of the Earth. Its length remains unchanged.

The definition of parallel transport is that the covariant derivative of a vector along the path is zero – the vector does not change direction along the path. This does not mean it will point in the same direction at either end of the path. The above example on the surface of a sphere is an example of parallel transport of a vector.

If the covariant derivative of a vector is zero, the Christoffel symbols are zero and so the coefficients of the 'extra bits of basis vectors' are zero.

With a little thought, the reader will see that the amount of change in the vectors direction depends upon the path it followed when it moved around the surface of the Earth. Different paths cause different changes. In curved spaces, the concept of two spatially separated vectors being parallel is meaningless because the route along which we would move one vector towards the other so we can compare them affects the moved vector; the direction of the moved vector is path dependent. However, the idea of keeping a vector pointing in the same direction as it is moved is meaningful and is called parallel transport of a vector.

If a vector is parallel along a curve, the curve is a geodesic of the space. In other words, if, at each point on the curve, the covariant derivative of the tangent vector is zero, the curve is a geodesic. Moving along such a curve corresponds to being in an inertial reference frame (constant velocity or free-fall in a gravitational field).

Parallel transport preserves the length of a vector and it also preserves the 2-dimensional angle between two vectors; in short, parallel transport preserves the 2-dimensional inner product of two vectors.

Intrinsic curvature can be defined by using parallel transport of a vector. We move a vector around a path and return the vector to its starting point thereby forming a complete circuit. Any change in orientation of the vector is a measure of the total intrinsic curvature embraced by the circuit. All there is to know of curvature is in the concepts of parallel transport around a circuit. Alternatively, the Riemann curvature tensor, R^a_{bcd}, (see later) is all there is to know about curvature. Without parallel transport (affine connection) it is meaningless to say that a space is curved or flat.

Of course, in our space-time, the change of orientation is measured as changes in 2-dimensional angles. We have two types of 2-dimensional rotations within our space-time and thus two types of change of orientation. We have three orthogonal copies of each type of 2-dimensional rotation in our space-time, and change of orientation is measured as six angles, three of each type. These six angles are the principal curvatures of the space.

Aside: Given that our space-time 'lost' its 4-dimensional rotation when it emerged from the super-imposition of the six A_3 algebras and has rotation within it only by the 'luck' of a coincidence of distance function with the 2-dimensional distance functions, it gives one pause for thought to consider whether or not a space with no sense of rotation could have any curvature or affine connection (parallel transport of vectors). Would a space without rotation be a space in any meaningful sense of the word?

Since tensors are products of vectors, if we can parallel transport vectors, we can parallel transport tensors.

The Riemann curvature tensor:
The reader is urged to think back to her days of elementary calculus when she was presented with a function as a graph and asked to find the maxima and minima of the function. By differentiating the function once and finding where the resulting differential was zero, she was able to find which points on the graph were horizontal. There

are maxima or minima at $\dfrac{dy}{dx} = 0$. However, to determine whether the horizontal points were maxima or minima, the reader had to take the second differential, $\dfrac{d^2x}{dy^2}$. Maxima or minima correspond to negative curvature or positive curvature as revisiting the graph will make clear. The important point is that the curvature is in the second differentials.

We are working in spaces which are locally flat everywhere. This means that, at any point in the space, the first differential of the metric tensor is always zero. We also have the fact that the second differentials determine the curvature.

Another way to look at curvature is to consider each of the, in 4-dimensions, 40 differentials of the ten components of the metric tensor:

$$\frac{\partial g_{mn}}{\partial x^r} = 0 \tag{8.5}$$

Expanding x^m to second order gives 40 equations of the form:

$$x^m = y^m + C^m_{nr} y^n y^r \tag{8.6}$$

Thus, all 40 of the first differential equations, (8.5), can be solved. This is another way of saying that the space is locally flat. Of course, $\dfrac{\partial V_m}{\partial x^r} = 0$ cannot be true globally in a particular curvi-linear coordinate system.

In 4-dimensions, there are ten possible second differentials of the metric tensor (ten for each component of the metric tensor but that is ten for the metric tensor taken as a whole):

$$\frac{\partial^2 g_{mn}}{\partial x^0 \partial x^0}, \ \frac{\partial^2 g_{mn}}{\partial x^0 \partial x^1}, \ \frac{\partial^2 g_{mn}}{\partial x^0 \partial x^2}, \ \frac{\partial^2 g_{mn}}{\partial x^0 \partial x^3}, \ \frac{\partial^2 g_{mn}}{\partial x^1 \partial x^1}$$
$$\frac{\partial^2 g_{mn}}{\partial x^1 \partial x^2}, \ \frac{\partial^2 g_{mn}}{\partial x^1 \partial x^3}, \ \frac{\partial^2 g_{mn}}{\partial x^2 \partial x^2}, \ \frac{\partial^2 g_{mn}}{\partial x^2 \partial x^3}, \ \frac{\partial^2 g_{mn}}{\partial x^3 \partial x^3} \tag{8.7}$$

If a space is curved in a particular way in one arbitrarily chosen set of co-ordinates, then it is equally curved in all co-ordinate systems. Thus, the expression for that curvature, which must be some function of the second differentials of the metric tensor, must be a tensor.

Remarkably, there is only one way the second differentials of the metric tensor can be combined to form a tensor. That unique combination of the second differentials of the metric tensor is called the Riemann curvature tensor. The Riemann curvature tensor is:

$$R^i_{jkl} = \frac{\partial \Gamma^i_{jl}}{\partial x^k} - \frac{\partial \Gamma^i_{jl}}{\partial x^l} + \Gamma^r_{jl}\Gamma^i_{rk} - \Gamma^r_{jk}\Gamma^i_{rl} \qquad (8.8)$$

The reader is reminded that the Christoffel symbols are functions of the metric, see (7.25). Perhaps of equal importance is the intimate connection between the curvature tensor and the affine connection.

The metric and the affine connection:

The curvature of a space arises from the connection which is the definition of parallel transport of a vector from the tangent space at one point to the tangent space at another point. There are many possible connections which can be defined upon a space, but the existence of a metric implies a unique particular one of those possible connections. If we use this unique connection, then we can take the view that the curvature of the space arises from the metric because the connection comes with the metric. Alternatively, we can take the view that the curvature of the space arises from the connection and the metric comes with the connection. This unique connection is:

$$\text{Torsoin Free} \qquad \Gamma^\lambda_{\mu\nu} - \Gamma^\lambda_{\nu\mu} = 0$$
$$\text{Everywhere Metric Compatible} \qquad D_\rho g_{\mu\nu} = 0 \qquad (8.9)$$

These two conditions single out the one unique torsion free metric compatible connection for any given metric.

This unique connection is called by several names:

a) The Levi-Civita connection

b) The Christoffel connection
c) The Riemann connection

As the reader has already guessed, GR uses the unique connection, and so, in GR, the curvature derives from the metric if you like, or, if you prefer, the curvature derives from the connection (which is what really happens).

If we had built a gravitational theory using a connection other than the unique Levi-Civita connection, then that theory would differ from GR by a tensor field. As such, by postulating another tensor field, we would be able to get GR from the theory. Who wants a tensor field that is surplus to requirements?

Back to the curvature tensor:
Using the unique Levi-Civita connection, we have, rewriting (8.8):

$$R_{ijkl} = \frac{1}{2}\left(\frac{\partial^2 g_{il}}{\partial x^j \partial x^k} + \frac{\partial^2 g_{jk}}{\partial x^i \partial x^l} - \frac{\partial^2 g_{ik}}{\partial x^j \partial x^l} - \frac{\partial^2 g_{jl}}{\partial x^i \partial x^k} \right)$$
$$+ \Gamma_{ilr}\Gamma^r_{jk} - \Gamma_{ikr}\Gamma^r_{jl} \qquad (8.10)$$

and:

$$R^\alpha_{\beta\mu\nu} = \frac{1}{2}g^{\alpha\sigma}\left(\frac{\partial^2 g_{\sigma\nu}}{\partial x^\beta \partial x^\mu} - \frac{\partial^2 g_{\sigma\mu}}{\partial x^\beta \partial x^\nu} + \frac{\partial^2 g_{\beta\mu}}{\partial x^\sigma \partial x^\nu} - \frac{\partial^2 g_{\beta\nu}}{\partial x^\sigma \partial x^\mu} \right) \qquad (8.11)$$

The equation (8.11) is not a tensor equation because it contains partial derivatives rather than covariant derivatives; it is true in only the co-ordinate system in which it was derived.

We emphasize that the Riemann curvature tensor is a function of the second differentials of the metric tensor. When we have the metric tensor, we know everything about the curvature of the space.

The Riemann curvature tensor can be defined in any space provided that space has within it an affine connection (a definition of what it means to parallel transport a vector from point to point).

Curvature is a tensor field. If the Riemann curvature tensor is zero, the space is flat.

The Riemann curvature tensor again:

The Riemann curvature tensor is related to the commutator of two covariant derivatives of vectors. We have:

$$D_s D_r V_n - D_r D_s V_n = R^t_{srn} V_t \qquad (8.12)$$

Partial derivatives commute in that the order in which we take two partial derivatives of a vector with respect to different variables does not affect the result. The order in which we take two covariant derivatives of a vector with respect to different variables does affect the result unless the space is flat. If the space is flat, the covariant derivative is the same thing as the partial derivative, and so the two covariant derivatives of a vector with respect to different variables will commute.

In many texts, very sensibly, the Riemann curvature tensor is defined by (8.12).

Note: the 2nd covariant derivatives of a scalar field do commute. For the scalar field, ϕ, we have:

$$D_\mu D_\nu \phi = D_\nu D_\mu \phi \qquad (8.13)$$

The non-commutative nature of the 2nd covariant derivatives of a vector field (and the commutative nature of the 2nd covariant derivatives of a scalar field) is invariant under change of co-ordinate system, of course.

Symmetries of the Riemann tensor:

The Riemann tensor R^a_{bcd} is known as a Riemann tensor of the 2nd kind. We can lower the upper index using the metric to produce the Riemann tensor of the 1st kind, R_{abcd}. The Riemann tensor has several symmetries. We have:

$$R_{abcd} = R_{cdab} = -R_{abdc} = -R_{bacd}$$
$$R_{abcd} + R_{acdb} + R_{adbc} = 0 \qquad (8.14)$$

The second of these identities is sometimes called the first Bianchi identity. There is another Bianchi identity, sometimes called the second Bianchi identity:

$$D_e R_{abcd} + D_c R_{abde} + D_d R_{abec} = 0 \qquad (8.15)$$

In other literature, we often find references to the Bianchi identity. These references always mean the second Bianchi identity, (8.15). The first Bianchi identity is not of immense importance, and it is often not mentioned. The Bianchi identities are differential equations.

In n dimensions, the Riemann curvature tensor has n^4 components, but, because of the symmetries within it, many of these components are not independent. In 2-dimensions, the Riemann curvature tensor has only one independent component, $\frac{1}{2} R_{\mu\nu}^{\mu\nu}$. In general, in n-dimensional space, there are $\frac{n^2(n^2-1)}{12}$ non-zero independent components of the Riemann tensor.

To give consistent results when calculating the amount of intrinsic curvature embraced within a circuit of a parallel transported vector, the Riemann curvature tensor must satisfy the first Bianchi identity. In other words, for curvature to make sense, the Riemann curvature tensor must satisfy the first Bianchi identity. The second Bianchi identity implies that the Einstein tensor has zero divergence. The zero divergence of the Einstein tensor is what pushes us towards the GR field equations. Zero divergence is associated with conservation laws.

Summary:
We have now covered Riemann geometry. Phew! The essential points of Riemann geometry are:

 a) Local flatness
 b) The Riemann curvature tensor
 c) The affine connection and covariant differentiation
 d) The metric tensor

The essential equations of Riemann geometry are:

$$\Gamma^{\lambda}_{\mu\nu} = \frac{1}{2}g^{\lambda\sigma}\left(\partial_{\mu}g_{\nu\sigma} + \partial_{\nu}g_{\sigma\mu} - \partial_{\sigma}g_{\mu\nu}\right) \tag{8.16}$$

$$D_{\mu}V^{\nu} = \partial_{\mu}V^{\nu} + \Gamma^{\nu}_{\mu\sigma}V^{\sigma} \tag{8.17}$$

$$R^{\rho}_{\sigma\mu\nu} = \partial_{\mu}\Gamma^{\rho}_{\nu\sigma} - \partial_{\nu}\Gamma^{\rho}_{\mu\sigma} + \Gamma^{\rho}_{\mu\lambda}\Gamma^{\lambda}_{\nu\sigma} - \Gamma^{\rho}_{\nu\lambda}\Gamma^{\lambda}_{\mu\sigma} \tag{8.18}$$

Chapter 9

Gravity

Principal curvatures:

We begin by raising an index of the Riemann tensor using the inverse metric tensor:

$$R^{\mu\nu}_{\rho\sigma} = g^{\nu\alpha} R^{\mu}_{\alpha\rho\sigma} \tag{9.1}$$

Now, we take any anti-symmetric tensor, $S^{\rho\sigma} = -S^{\sigma\rho}$, and we form the product of these two tensors, $R^{\mu\nu}_{\rho\sigma} S^{\rho\sigma}$. This product is an anti-symmetric tensor. We then seek to solve the eigenvalue equation:

$$\frac{1}{2} R^{\mu\nu}_{\rho\sigma} S^{\rho\sigma} = \kappa S^{\mu\nu} \tag{9.2}$$

wherein κ is a real number. We have here six linear homogeneous equations with six unknowns. (A 4-dimensional anti-symmetric tensor has six independent non-zero components; the diagonal components are zero.) In most cases, such a set of equations would have only trivial $S^{\rho\sigma} = 0$ solutions; however, there are non-trivial solutions if the determinant of the coefficients is zero. The determinant is a 6th degree polynomial, and the six solutions are the six principal curvatures. These are six real numbers which are independent of the co-ordinate system and express the curvature at any point in the space.

The six principal curvatures in space-time are a measure of the strength of the gravitational field at the given point.

Aside: The reader might notice the coincidence between the six A_3 algebras and the six principal curvatures. At this time, it is thought that there is no significance in this coincidence.

Tidal forces and curvature:

A uniform gravitational field does not exist in nature, but it is approximated by the surface of a large planet. In a uniform gravitational field, the field lines are exactly parallel in a Euclidean sense and the gravitational field has the same strength at every point in space. The field lines of the gravitational field of the Earth are like a spiky ball; they radiate outward from the centre of the Earth. As we move away from the centre of the Earth, the gravitational field lines become less dense and so the gravitational field weakens.

Let us imagine a painter falling in a falling lift-shaft. A long way from the Earth, the painter takes two paint brushes and puts a paint brush at either side of the lift-shaft. The paint brushes, like the painter, are in free-fall and so they stay where they were put as they fall down along a gravitational field line towards the centre of the Earth. After a while, the painter notices that the two brushes are closer together than they were initially. As she watches, she sees the brushes slowly moving towards each other as if they were attracted to each other by some unseen force. What is happening is, as each brush follows a gravitational field line toward the centre of the Earth, they move closer together as the radial field lines become closer together. The force pushing the brushes together is a gravitational tidal force.

In a non-uniform gravitational field, we have gravitational tidal forces. We do not have gravitational tidal forces in a uniform gravitational field (parallel field lines). Since uniform gravitational fields have no tidal forces, the associated space-time is without curvature. It is non-uniform gravitational fields which are associated with curvature. Uniform acceleration can be transformed away by a change of co-ordinates; tidal forces cannot be transformed away by a change of co-ordinates. In other words, uniform acceleration is just a result of the co-ordinate system we chose, but tidal forces are real. To say uniform acceleration is just a result of our choice of co-ordinates is to say that an observer in free-fall is in an inertial reference frame.

The curvature tensor is related to tidal forces. Imagine an initially flat rubber disc initially on a flat plane having to slide over humps and hollows as it moves around but having to stay within the 2-dimensional space. The rubber disc will be stretched and distorted as it moves over the humps and hollows. There are forces associated with distorting the rubber disc; these are tidal forces.

The Ricci tensor and the Einstein tensor:
There is a tensor of great importance in GR called the Ricci tensor. The Ricci tensor is named after Gregorio Ricci-Curbastro (1853-1925). It is used to form the Einstein tensor. The Ricci tensor is formed by contracting the Riemann curvature tensor on the first and third indices. Other contractions of the Riemann curvature tensor could be taken, but, because the Riemann curvature tensor is anti-symmetric on the first and second indices and anti-symmetric on the third and fourth indices, all the other possible contractions are either zero or the same as the Ricci tensor except for the sign[35]. We have:

$$R_{ijk}^{k} = R_{ij} = \frac{\partial \Gamma_{ik}^{k}}{\partial x^{j}} - \frac{\partial \Gamma_{ij}^{k}}{\partial x^{k}} + \Gamma_{ik}^{r}\Gamma_{rj}^{k} - \Gamma_{ij}^{r}\Gamma_{rk}^{k} \qquad (9.3)$$

The Ricci tensor is symmetric.

$$R_{ab} = R_{ba} \qquad (9.4)$$

Although the Ricci tensor is, in general, not zero, the Ricci tensor is zero in empty space. We have:

$$_{Empty\ Space}R_{ab} = 0 \qquad (9.5)$$

This is ten equations. It is not sixteen because the Ricci tensor is symmetric. Moving from an empty space to a space with mass is moving from the Ricci tensor being zero to the Ricci tensor not being zero.

The Ricci tensor is the trace of the Riemann curvature tensor, and so not all components of the Riemann curvature tensor are included in the Ricci tensor. The components of the Riemann curvature tensor which are not included in the Ricci tensor together form a tensor called the Weyl tensor. The Weyl tensor is of no interest to us.

---

[35] The reader might meet other literature which uses the negative of the Ricci tensor.

We can raise an index on the Ricci tensor:

$$R^i_j = g^{ik} R_{kj} \tag{9.6}$$

Having raised the index, we can contract the Ricci tensor to give a curvature invariant known as the Ricci scalar, R.

From the Ricci tensor, the metric tensor, and the Ricci scalar, we can form the Einstein tensor:

$$G^{\mu\nu} = R^{\mu\nu} - \frac{1}{2} g^{\mu\nu} R \tag{9.7}$$

The second Bianchi identity says that the covariant derivative of this is zero. We have:

$$D_\nu \left(R^{\mu\nu} - \frac{1}{2} g^{\mu\nu} R \right) = 0$$
$$D_\nu G^{\mu\nu} = \partial_\nu G^{\mu\nu} + \Gamma^\mu_{\alpha\nu} G^{\alpha\nu} + \Gamma^\nu_{\alpha\nu} G^{\alpha\mu} = 0 \tag{9.8}$$

And so the covariant divergence of the Einstein tensor is zero at all points in a space. As we mentioned earlier, it is the zero divergence of the Einstein tensor that led Einstein to put the Einstein tensor equal to the energy-momentum tensor.

The energy-momentum tensor:
The energy-momentum tensor is also called the stress–energy tensor. It is a 4-dimensional second rank contravariant tensor denoted by $T^{\mu\nu}$. Some of the components of this tensor are energy and some of the components are momentum. In GR, the energy-momentum tensor takes the place of mass in the Newtonian gravity theory. The energy-momentum tensor is seen as the origin of gravity in a way very similar to the way mass is seen as the origin of gravity in Newtonian gravity.

The covariant divergence of the energy-momentum tensor is zero implying the conservation of energy, mass, momentum and angular momentum. The covariant divergence of the energy-momentum tensor is:

$$D_v T^{\mu\nu} = \partial_v T^{\mu\nu} + \Gamma^{\mu}_{\alpha v} T^{\alpha\nu} + \Gamma^{v}_{\alpha v} T^{\alpha\mu} = 0 \qquad (9.9)$$

The Γ^{a}_{bc} terms in the covariant derivative are the terms associated with gravity. The reader is urged to compare this to the divergence of the Einstein tensor, (9.8).

The GR field equations:
We have, because of the second Bianchi identity, the fact that the Einstein tensor has zero covariant divergence. We have that the energy-momentum tensor has zero covariant divergence. Putting, for no reason other than we do not know what else to do, the Einstein tensor equal to the energy-momentum tensor, $T^{\mu\nu}$. Gives the field equations of GR:

$$R^{\mu\nu} - \frac{1}{2} g^{\mu\nu} R = 8\pi G T^{\mu\nu} \qquad (9.10)$$

This is an equation between 4×4 matrices. The left-hand side of this measures curvature and is called the Einstein tensor, $G^{\mu\nu}$. The GR field equations are sometimes written as:

$$G^{\mu\nu} = 8\pi G T^{\mu\nu} \qquad (9.11)$$

We have included a proportionality constant, $8\pi G$ where G is Newton's gravitational constant. This constant is found by equating the GR result to the Newtonian result in the weak field limit. We can think of $G^{\mu\nu}$ as some of the components of the curvature tensor.

The GR field equations are ten independent equations between the components of the tensors (elements of the matrices) on each side of the equations. There are ten equations rather than sixteen because the tensors are symmetric. Further the Bianchi identity, $D^{\mu} G_{\mu\nu} = 0$ constrains the Ricci tensor in four ways, and so there are really only six truly independent equations in the GR field equations, (9.10).

Within Newtonian gravity theory, the field equation is:

$$\nabla^2 \Phi = 4\pi G \rho \qquad (9.12)$$

91

Where Φ is the gravitational potential and ρ is the mass-density. This is equivalent to:

$$F = -\frac{GMm}{r^2} \tag{9.13}$$

and so, in GR, we have (9.10) instead of the Newtonian (9.13).

The GR field equations fit observation as perfectly as we are able to measure. Of course, the field equations of any theory of gravity should reduce to the field equations of Newtonian gravity in the weak field limit; these equations do that.

There is only one field equation in Newtonian gravity. There are four field equations in Classical electro-magnetism. In GR, there are ten field equations; (9.10) & (9.11) are ten equations; there is one equation for each of the ten independent components of the tensor. The ten field equations determine the ten components, $g_{\mu\nu}$, of the metric tensor. Perhaps it would be more instructive to rewrite (9.10) as:

$$g^{\mu\nu} = -\frac{2}{R}\left(8\pi G T^{\mu\nu} + R^{\mu\nu}\right) \tag{9.14}$$

Newtonian gravity has a scalar potential. Electromagnetism has a 4-vector potential. GR has a tensor potential (matrix potential), which is the metric tensor, $g_{\mu\nu}$. It is because GR has a tensor potential that the postulated gravitons of postulated quantum gravity are spin 2 bosons.

GR does not treat the gravitational field as a source of gravity. This is why two masses bound together by gravity are of the same total mass as would be the case if the masses were separate. This is not the case for an atomic nucleus where the mass of the nucleus is less than the total of the separate particles; the difference being binding energy. Gravity does couple, is affected by, the binding energy between sub-atomic particles.

Although gravity does not treat the gravitational field as a source of gravity, the gravitational field does affect (couple to) itself. This is the cause of the GR precession of the orbit of the planet Mercury. It is also the reason why the gravitational field equations are non-linear.

Empty Space:

In the absence of matter, the Ricci tensor is zero:

$$R^{\mu\nu} = 0 \tag{9.15}$$

This implies that the Ricci scalar is zero, $R = 0$, and hence that the Einstein tensor is zero:

$$R^{\mu\nu} - \frac{1}{2} g^{\mu\nu} R = 0 \tag{9.16}$$

This is the GR field equation of empty space.

In the presence of matter, the Einstein tensor is not zero. We put it equal to another tensor:

$$R^{\mu\nu} - \frac{1}{2} g^{\mu\nu} R = Y^{\mu\nu} \tag{9.17}$$

The 2nd Bianchi identity tells that:

$$D_\nu \left(R^{\mu\nu} - \frac{1}{2} g^{\mu\nu} R \right) = 0 \tag{9.18}$$

This requires:

$$D_\nu Y^{\mu\nu} = 0 \tag{9.19}$$

Conservation of energy and momentum leads us to assume that the tensor $Y^{\mu\nu}$ is the energy-momentum tensor.

Since geodesics, like the orbits of planets, are determined by the metric, it seems reasonable that the metric should be determined by the sources of the field, and so it seems reasonable to put the Einstein tensor equal to the energy-momentum tensor.

Aside: The energy-momentum tensor of electromagnetism is[36]:

$$T^{\mu\nu}_{EMag} = F^{\mu\lambda} F^\nu_\lambda - \frac{1}{4} \eta^{\mu\nu} F^{\lambda\sigma} F_{\lambda\sigma} \tag{9.20}$$

---

[36] See: Sean M. Carroll Space-time and Geometry pg 44

We note that $F^{\mu\lambda}F^{\nu}_{\lambda}$ is just a multiple of $F^{\mu\nu}$ and $\eta^{\mu\nu} = g^{\mu\nu}$ in flat space-time and $F^{\lambda\sigma}F_{\lambda\sigma}$ is just a real number. We see a remarkable similarity to the field equations of GR, (9.10). Later in this book, we show both GR and electromagnetism emerging alongside each other from the super-imposition of the A_3 algebras.

The equations of motion:

An observer in free-fall is an inertial observer. From special relativity, we have the equations of motion:

$$\frac{d^2\xi}{d\tau^2} = 0 \quad \& \quad c^2 d\tau^2 = \eta_{\alpha\beta}d\xi^{\alpha}d\xi^{\beta} \qquad (9.21)$$

where $\eta_{\alpha\beta}$ is the flat metric tensor $diag(1,-1,-1,-1)$. Transforming this to another set of co-ordinates, x^{μ}, gives:

$$d\xi^{\mu} = \frac{\partial\xi^{\mu}}{\partial x^{\nu}}dx^{\nu} \qquad (9.22)$$

Substituting this into the special relativity equations of motion (9.21) gives:

$$\frac{d^2x^{\mu}}{d\tau^2} + \Gamma^{\mu}_{\alpha\beta}\frac{dx^{\alpha}}{d\tau}\frac{dx^{\beta}}{d\tau} = 0$$
$$c^2 d\tau^2 = g_{\alpha\beta}dx^{\alpha}dx^{\beta} \qquad (9.23)$$

These are the two dynamical equations of GR. The first equation of (9.23) is known as the geodesic equation; it says that objects will move along the straightest possible route.

Aside: The reader might see the geodesic equation written with a parameter, λ, along a curve in place of τ as:

$$\frac{d^2x^{\mu}}{d\lambda^2} + \Gamma^{\mu}_{\alpha\beta}\frac{dx^{\alpha}}{d\lambda}\frac{dx^{\beta}}{d\lambda} = 0 \qquad (9.24)$$

> In this case, we say that the parameterized curve $x^\mu(\lambda)$ is a geodesic if it satisfies (9.24).

The Newtonian dynamical equation is:

$$a = -\nabla\Phi \qquad or \qquad F = ma \qquad (9.25)$$

The story so far:

We now have the field equations part of GR. We have taken many chapters to get here. The other part of GR is the equivalence principle and its consequences. We do this second part in the single following chapter; it is much easier than the field equations.

Aside: GR in 2 & 3 dimensions:

Remarkably, there is no gravity based on curvature in 3-dimensional Riemann space-time. In only 3-dimensional space-time (2 space dimensions and 1 time dimension), we have the Riemann curvature tensor in terms of the Ricci tensor:

$$R_{\alpha\beta\gamma\delta} = R_{\alpha\gamma}g_{\beta\delta} + R_{\beta\delta}g_{\alpha\gamma} - R_{\alpha\delta}g_{\beta\gamma} - R_{\beta\gamma}g_{\alpha\delta}$$
$$-\frac{1}{2}g^{\mu\nu}R_{\mu\nu}\left(g_{\alpha\gamma}g_{\beta\delta} - g_{\alpha\delta}g_{\beta\gamma}\right) \qquad (9.26)$$

The term $g^{\mu\nu}R_{\mu\nu}$ is the 3-dimensional Ricci scalar. Thus, in 3-dimensional space-time, if $R_{ab} = 0$ then $R_{abcd} = 0$ which says that empty 3-dimensional space-time has to be flat – no GR type gravity.

In 2-dimensional space-time, curvature is a single real number, and so GR type gravity in 2-dimensional space-time is quite simple.

We see that only if our space-time has four or more dimensions do we have meaningful GR type of gravity associated with curvature.

Chapter 10

The Equivalence Principle

The theory of GR has two disconnected parts. One of these parts is the field equations. The other part is the phenomena deduced from the equivalence principle. In this one short chapter, we make those deductions.

The equivalence principle is an assertion made originally by Einstein. The assertion seems to fit our observed universe so well that we take it to by a blindingly obvious statement of fact. The equivalence principle says that all inertial observers will observe the same physics. Put another way, the laws of physics are the same in all inertial frames of reference. The equivalence principle implies the equality of gravitational and inertial mass, and the equality of gravitational and inertial mass implies the equivalence principle.

There is a little bit of a shock when we realise that we have to count observers who are in free-fall within a gravitational field as observers in an inertial reference frame. It was Einstein's realisation that a painter falling from a ladder was an inertial observer until he hit the floor that led to the development of the theory of general relativity.

An observer standing on the surface of the Earth is not an inertial observer. An observer standing on the surface of the Earth is an accelerated observer; she can feel the acceleration pressing on her feet; she can test her status by dropping her hat and watching it fall.

An observer falling from a plane toward the Earth's surface is an inertial observer because there is no acceleration pressure on that observer. Of course, a planet in orbit around the sun is also in free-fall.

Let us put our painter in a falling lift within a lift shaft with a pulsing laser beam pointing horizontally across the lift. To the inertial observer, the laser beam crosses the lift horizontally maintaining its distance from the floor of the lift. Of course it does; to an inertial observer, light travels in straight lines. To the accelerated observer on the Earth's surface, by the time the pulse of light hits the far wall of the lift, the lift has fallen a little. However, both observers must observe the same event of the light pulse hitting the wall of the lift at the same point, and so, to the accelerated observer, the pulse of light must travel on a curved path.

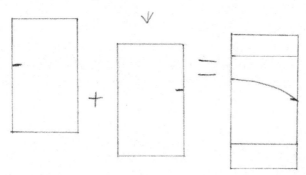

To an accelerated observer, light bends in its path. Gravity bends light.

The bending of light as it passes by the sun has been observed using radio waves. Radio waves are more suitable than optical frequencies for the experimentalist. Astronomers routinely observe 'Einstein rings' that are produced by this effect.

There is a further effect perceived by the accelerated observer on the Earth's surface. The path of the light ray is longer than it appears to the inertial observer; but light travels at the same speed to all observers; that's a law of physics. If light is to travel the curved path in the same time for both observers, the accelerated observer must perceive some time dilation relative to the inertial observer. This is not time dilation due to relative velocity. In a falling lift, there will be time dilation due to relative velocity, but the accelerated observer sees an additional time dilation due to his own acceleration. The reader might well be familiar with the Feynman clock on a moving train leading to time dilation. We have above a Feynman clock in what appears to the observer on the Earth's surface to be an accelerating train. The acceleration curves the path of the light; pure relative motion would give a straight line at an angle to the horizontal.

To an observer on the Earth's surface, it seems that time slows down in a gravitational field. The stronger the gravitational field, the greater the acceleration of the observer on the planet's surface, and the more time dilation there is due to gravity. Because the gravitational field of the Earth is weaker further from the centre of the Earth, clocks run slower at the Earth's surface than they do at high altitude. The global positioning system, GPS, has to take account of this to work accurately. Clocks at the Royal Greenwich Observatory at an altitude of 80 feet above sea level lose 5 microseconds per year compared with clocks at the National Bureau of Standards at Boulder, Colorado at an altitude of 5,400 feet above sea level. We see this difference because we are accelerated observers.

Gravitational red-shift:
The inertial observer in the free-falling lift emits a photon of light from the floor of the lift towards the ceiling of the lift. In the eyes of the inertial observer, the photon travels to the ceiling with no change to its frequency.

The accelerated observer standing on the Earth's surface sees the photon initially Doppler shifted by an amount due to the velocity of the lift when the photon left the floor of the lift and Doppler shifted by a larger amount due to the higher velocity of the lift when the photon hits the ceiling of the lift. To the accelerated observer, the frequency of the photon has changed as the photon climbed the height of the lift. The amount of the change is proportional to the acceleration of the lift which is the strength of the gravitational field. This change of frequency to photons climbing out of a gravitational field is called the gravitational redshift.

The change in frequency due to gravitational red-shift corresponds to the appropriate amount of gravitational time dilation, and we could have deduced the time dilation from the change of frequency.

Experiments by Pound & Rebka in 1960 observed this predicted gravitational red-shift. The result was confirmed to greater accuracy in 1965 by Pound and Snider.

Geodesic world lines:
When we take an observer in free-fall to be in an inertial reference frame, we are effectively saying that the observer is travelling along

a geodesic through space-time. In the same way that a planet in orbit around a star is if free-fall and thereby travels along a geodesic in space-time, so all inertial observers travel along geodesics in space-time. Deviation from a geodesic is associated with a force.

Torsion and the equivalence principle:
The equivalence principle asserts that there is zero torsion, but torsion is not ruled out by observation.

Summary:
From nothing more than the physical equivalence of all inertial reference frames and that a free-falling object is in an inertial reference frame, we have deduced gravitational red-shift, gravitational time dilation, the bending of light by gravity, and that inertial observers move along a geodesic in space-time.

If we define a straight line through space to be the path followed by light, we have deduced curved space-time.

Chapter 11

Other Theories of Gravity

Although GR is a perfectly successful theory as judged by observation, there are aspects of it that lead theorists to opine that it needs to either be developed or replaced with some other theory. There seems to be no place in GR for Plancks constant and hence no place for a quantum theory of gravity. The field equations of GR are not deduced but guessed; perhaps other field equations might fit observation just as well. For these reasons alternative theories of gravity have been considered.

Flat space theories:
The first type of alternative theory is to reject the idea of space-time curvature and to try to describe gravity with a type of field over space-time rather than within it. Such a field could be either a scalar field or a vector field or a tensor field. Such theories have been proposed[37] but they do not fit observation. Scalar and vector field theories do not predict the bending of light. The tensor field theory predicts the bending of light but gives a precession of the planetary orbits that is too large. All of these theories, which, because they reject curvature, are known as flat space theories, fail to predict gravitational time dilation.

Cartan's torsion theory:
Another alternative theory was developed by Elie Cartan (1896-1951) in which the Christoffel symbols are not symmetric in the lower indices. This generates a torsion tensor:

$$S^{\alpha}_{\beta\lambda} = \Gamma^{\alpha}_{\beta\lambda} - \Gamma^{\alpha}_{\lambda\beta} \qquad (11.1)$$

---

[37] See Gravitation by Misner Thorne & Wheeler Chap. 7

Although the equivalence principle requires that the Christoffel symbols are symmetric in the lower indices, a symmetric connection, the dynamic equations of gravity are unaffected if the Christoffel symbols are anti-symmetric in the lower indices, and so we cannot dismiss Cartan's torsion theory by observation.

The Brans-Dicke theory:

The Brans-Dicke theory of gravity is also called the scalar-tensor theory of gravity. In this theory, the gravitational constant, G, is taken to be a scalar field, ϕ, which varies over space-time. Other than this, the theory matches GR in every way.

Gravito-electromagnetism:

There is a somewhat esoteric area of physics called gravito-electromagnetism, GEM[38]. GEM is concerned, along with other things, with the Lense-Thirring effect presently being tested by the satellite Gravity Probe B (also called the Stanford Gyroscope Experiment). GEM is thought to be connected to quantum gravity. GEM is often, in the opinion of some wrongly[39], referred to as 'Frame Dragging'. GEM is developed as an analogy to electromagnetism that *"...supplies us with a familiar model for the extra gravitational force created by the motion of the sources and felt only by moving particles..."*[40].

Traditionally, GEM is seen as an approximation to general relativity and is based upon linearized GR equations, however, it can be treated exactly. The exact treatment takes a vector potential, A_μ, and defines two sectors of its first derivatives as:

$$F_{\mu\nu} = \partial_\mu A_\nu - \partial_\nu A_\mu$$
$$H_{\mu\nu} = \partial_\mu A_\nu + \partial_\nu A_\mu$$

(11.2)

---

[38] A summary of GEM is offered in the paper Gravitoelectromagnetism: A Brief Review, Bahram Mashhoon arXiv:gr-qc/0311030v2 17th April 2008.
[39] W. Rindler, The case against space dragging. Phys. Lett. A233, 25 (1997)
[40] W. Rindler, Relativity, Special, General and Cosmological. Pg 336. Oxford Uni. Press 0-19-850836-0

In GEM, there are two fields analogous to the electric and magnetic fields of electromagnetism. There are also GEM Maxwell equations. We list them alongside the electromagnetic Maxwell equations:

$$\nabla \cdot E_g = -4\pi G \rho_g \qquad\qquad \nabla \cdot E = \frac{1}{\varepsilon_0}\rho$$

$$\nabla \cdot B_g = 0 \qquad\qquad \nabla \cdot B = 0$$

$$\nabla \times E_g = -\frac{\partial B_g}{\partial t} \qquad\qquad \nabla \times E = -\frac{\partial B}{\partial t}$$

$$\nabla \times B_g = 4\left(-\frac{4\pi G}{c^2}J_g + \frac{1}{c^2}\frac{\partial E_g}{\partial t}\right) \qquad \nabla \times B = \frac{1}{\varepsilon_0 c^2}J + \frac{1}{c^2}\frac{\partial E}{\partial t}$$

$$(11.3)$$

We will meet something very much like GEM later in this book. It might be, with a few sign changes, that GEM is quantum gravity. If it is so, then we have three copies of quantum gravity, one for each pair of A_3 algebras.

Comments:

Like GR, none of the alternative theories answer questions like why space-time is 4-dimensional, why we have the observed distance function, or why we see two types of 2-dimensional rotations within our 4-dimensional space-time. Nor do they provide an explanation for the expansion of the universe. We, at present, have no theory which predicts the observed rotation curves of galaxies which we associate with some kind of 'dark matter'.

Chapter 12

From Spinor Algebras to Space-time

Within this chapter, we again examine how Riemann space emerges from the spinor algebras. We repeat some of the material we presented in the earlier chapter upon this subject, but we add in much more material to broaden the reader's understanding. Having worked through the details of Riemann geometry, the reader might look upon the repeated and extended material from a better informed perspective.

Spinor algebras are just the many different forms of complex numbers. There are two types of 2-dimensional complex numbers, the Euclidian complex numbers, $\mathbb{C}$, with which the reader is doubtless familiar as the 2-dimensional flat plane, and the hyperbolic complex numbers, $\mathbb{S}$, which are the 2-dimensional space-time of special relativity. These two algebras are based on the finite group C_2. There are four types of 3-dimensional complex numbers based on the finite group C_3; two of these are algebraically isomorphic. There are three copies of eight types of 4-dimensional complex numbers based on the finite group C_4; many of these are algebraically isomorphic. There are eight commutative types of 4-dimensional complex numbers based on the finite group $C_2 \times C_2$, and there are eight non-commutative types of 4-dimensional complex numbers based on the finite group $C_2 \times C_2$. Similarly, there are higher dimensional complex numbers based on all the higher order finite groups.

The spinor algebras of interest:
Our interest is with only the two types of commutative 2-dimensional complex numbers from the commutative group C_2 and the eight non-commutative types of 4-dimensional complex numbers from the commutative group $C_2 \times C_2$. It is remarkable that

103

a commutative group should hold within it non-commutative algebras[41]. Other than the 2-dimensional algebras, the commutative algebras seem to play no part in the physical universe.

The two 2-dimensional algebras are:

$$S = \exp\left(\begin{bmatrix} t & z \\ z & t \end{bmatrix}\right) = \begin{bmatrix} e' & 0 \\ 0 & e' \end{bmatrix} \begin{bmatrix} \cosh z & \sinh z \\ \sinh z & \cosh z \end{bmatrix} \tag{12.1}$$

$$C = \exp\left(\begin{bmatrix} x & y \\ -y & x \end{bmatrix}\right) = \begin{bmatrix} e^x & 0 \\ 0 & e^x \end{bmatrix} \begin{bmatrix} \cos y & \sin y \\ -\sin y & \cos y \end{bmatrix} \tag{12.2}$$

Of course, the hyperbolic rotation matrix is just the Lorentz transformation of special relativity:

$$\begin{bmatrix} \cosh z & \sinh z \\ \sinh z & \cosh z \end{bmatrix} = \begin{bmatrix} \gamma & v\gamma \\ v\gamma & \gamma \end{bmatrix} \tag{12.3}$$

Of the eight non-commutative 4-dimensional algebras, two are quaternion type algebras, the quaternions and the anti-quaternions:

$$\mathbb{H} = \begin{bmatrix} a & b & c & d \\ -b & a & -d & c \\ -c & d & a & -b \\ -d & -c & b & a \end{bmatrix} \qquad \mathbb{H}_{Anti} = \begin{bmatrix} a & b & c & d \\ -b & a & d & -c \\ -c & -d & a & b \\ -d & c & -b & a \end{bmatrix} \tag{12.4}$$

These are algebraically isomorphic but are written in two different bases. The commutation relations of the anti-quaternions are the reverse of the commutation relations of the quaternions. These algebras are the spinor algebra forms of the Clifford algebra $Cl_{0,2}$.

The other six non-commutative 4-dimensional spinor algebras are the A_3 algebras which we present again for the convenience of the reader:

---

[41] See Dennis Morris: The Physics of Empty Space.

$$SSA^* = \exp\left(\begin{bmatrix} a & b & c & d \\ b & a & -d & -c \\ c & d & a & b \\ -d & -c & b & a \end{bmatrix}\right) \qquad SSA^*_{Anti} = \exp\left(\begin{bmatrix} a & b & c & d \\ b & a & d & c \\ c & -d & a & -b \\ -d & c & -b & a \end{bmatrix}\right)$$

$$(12.5)$$

$$SAS = \exp\left(\begin{bmatrix} a & b & c & d \\ b & a & d & c \\ -c & d & a & -b \\ d & -c & -b & a \end{bmatrix}\right) \qquad SAS_{Anti} = \exp\left(\begin{bmatrix} a & b & c & d \\ b & a & -d & -c \\ -c & -d & a & b \\ d & c & b & a \end{bmatrix}\right)$$

$$(12.6)$$

$$ASS = \exp\left(\begin{bmatrix} a & b & c & d \\ -b & a & -d & c \\ c & -d & a & -b \\ d & c & b & a \end{bmatrix}\right) \qquad ASS_{Anti} = \exp\left(\begin{bmatrix} a & b & c & d \\ -b & a & d & -c \\ c & d & a & b \\ d & -c & -b & a \end{bmatrix}\right)$$

$$(12.7)$$

These are algebraically isomorphic but are written in six different bases. They are the spinor algebra forms of the Clifford algebra $Cl_{2,0} \cong Cl_{1,1}$. All eight of these non-commutative algebras have the general form:

$$\begin{bmatrix} a & b & c & d \\ \beta b & a & \dfrac{\beta}{\varepsilon}d & \varepsilon c \\ \eta c & -\dfrac{\eta}{\varepsilon}d & a & -\varepsilon b \\ -\dfrac{\beta\eta}{\varepsilon^2}d & \dfrac{\eta}{\varepsilon}c & -\dfrac{\beta}{\varepsilon}b & a \end{bmatrix} \qquad (12.8)$$

This is a quaternion when $\{\beta < 0,\ \varepsilon > 0,\ \eta < 0\}$. We usually set the parameters $\{\beta,\varepsilon,\eta\}$ equal to ± 1 for convenience. It is important to realise that the number and bases of spinor algebras arise not by human choice but from the mathematics that leads to (12.8).

Spinor rotations and space-time rotations:
There are two types of rotations in the universe. There is the 2-dimensional rotations about an axis (or about two axes) with which we are experientially accustomed in our 4-dimensional space-time. Leonard Euler (1707-1783) proved that all rotations in the 3-dimensional spatial part of our space-time are 2-dimensional and are about an axis – he got that right. There are also spinor rotations which are rotations about no axis and are multi-angular. The intrinsic spin of the electron is associated with a spinor rotation whereas the orbital rotation of a planet moving around the sun is associated with the Euler type of rotation. Euler knew nothing of spinor rotation; he had only one eye on what he was doing.

An example of a spinor rotation is rotation in the complex plane. The complex plane is 2-dimensional; there is no third dimension to be the axis of the rotation. The complex plane rotation matrix is as shown above, (12.2); it is a 2-dimensional rotation in a 2-dimensional space. The equality of the dimension of the space and the dimension of the rotation ensures that there can be no spare dimension unaffected by the rotation to be an axis of rotation. Rotation in the complex plane is not a 'double cover' rotation.

Another example of a spinor rotation is the quaternion rotation in quaternion space:

$$\mathbb{H}_{Rot} = \exp\left(\begin{bmatrix} 0 & b & c & d \\ -b & 0 & -d & c \\ -c & d & 0 & -b \\ -d & -c & b & 0 \end{bmatrix}\right) \qquad (12.9)$$

$$\mathbb{H}_{Rot} = \begin{bmatrix} \cos(\lambda) & \dfrac{b}{\lambda}\sin(\lambda) & \dfrac{c}{\lambda}\sin(\lambda) & \dfrac{d}{\lambda}\sin(\lambda) \\[2mm] -\dfrac{b}{\lambda}\sin(\lambda) & \cos(\lambda) & -\dfrac{d}{\lambda}\sin(\lambda) & \dfrac{c}{\lambda}\sin(\lambda) \\[2mm] -\dfrac{c}{\lambda}\sin(\lambda) & \dfrac{d}{\lambda}\sin(\lambda) & \cos(\lambda) & -\dfrac{b}{\lambda}\sin(\lambda) \\[2mm] -\dfrac{d}{\lambda}\sin(\lambda) & -\dfrac{c}{\lambda}\sin(\lambda) & \dfrac{b}{\lambda}\sin(\lambda) & \cos(\lambda) \end{bmatrix} \qquad (12.10)$$

$$\lambda = \sqrt{b^2 + c^2 + d^2} \qquad (12.11)$$

Notice that the quaternion trigonometric functions accept three arguments (angles) each. This is very different from three 2-dimensional rotations accepting only one angle each. Again, this is not rotation about an axis; the multi-angular nature of spinor rotation is seen more easily in the quaternions than in the 2-dimensional rotations. Quaternion rotation is also 'double cover' rotation. The square root sign causes it.[42]

Spinor rotations are spinor algebra rotations. Every spinor algebra has a polar form containing a rotation matrix. The rotation matrix is the spinor rotation. This is why we refer to complex number algebras as spinor algebras.

The 3-dimensional rotation matrix that represents spatial rotation in our space-time is of the form:

$$\begin{bmatrix} \cos\theta & \sin\theta & 0 \\ -\sin\theta & \cos\theta & 0 \\ 0 & 0 & 1 \end{bmatrix} \qquad (12.12)$$

The eigenvectors and eigenvalues of this rotation matrix are:

---

[42] See: Dennis Morris The Naked Spinor.

$$(\cos\theta + i\sin\theta)\begin{bmatrix} -i \\ 1 \\ 0 \end{bmatrix}, \quad (\cos\theta - i\sin\theta)\begin{bmatrix} i \\ 1 \\ 0 \end{bmatrix}, \quad 1\begin{bmatrix} 0 \\ 0 \\ 1 \end{bmatrix}$$

$$(12.13)$$

We see that there is one eigenvector which is independent of the rotation angle θ. This is the axis of rotation. The eigenvectors and eigenvalues of the spinor rotation matrix:

$$\begin{bmatrix} \cos\theta & \sin\theta \\ -\sin\theta & \cos\theta \end{bmatrix}$$

$$(12.14)$$

are:

$$(\cos\theta + i\sin\theta)\begin{bmatrix} -i \\ 1 \end{bmatrix}, \quad (\cos\theta - i\sin\theta)\begin{bmatrix} i \\ 1 \end{bmatrix} \qquad (12.15)$$

We see there are no eigenvectors that are independent of the rotation angle θ. This is not rotation about an axis. Of course, we should really have used a 4×4 matrix to represent the spatial rotation (12.12) in our 4-dimensional space-time. If we had done this, we would have found two eigenvectors that are independent of the rotation angle θ, and so 2-dimensional rotation in our 4-dimensional space-time is really rotation about two axes.

Fabrication from 2-dimensional spinor algebras:
We have only two 2-dimensional spinor rotations which survive super-imposition. These are the Euclidean rotation of the 2-dimensional complex numbers, $\mathbb{C}$, which preserves the distance function $d^2 = x^2 + y^2$ and the space-time rotation of the hyperbolic complex numbers, $\mathbb{S}$, which preserves the distance function $d^2 = t^2 - z^2$. These are all there are from which to fabricate a higher dimensional emergent expectation space such as our 4-dimensional space-time.

If we are able to fabricate a higher dimensional emergent expectation space from these 2-dimensional rotations, then the distance function of that higher dimensional emergent expectation space must accommodate the 2-dimensional distance functions. Thus, the distance function of the higher dimensional emergent expectation space must be a quadratic form with as many variables as the dimension of the emergent expectation space. The quadratic form will have a number of plus signs and a number of minus signs; these numbers will dictate how many 2-dimensional planes (pairs of variables) have Euclidean rotation and how many 2-dimensional planes (pairs of variables) have space-time rotation (Lorentz boost).

In 2-dimensions, we have four possible ways to put two variables into a quadratic form distance function. The two squares of the variables can either have the same sign, which is the conserved distance function of a Euclidean rotation, or the two squares of the variables can have different signs, which is the conserved distance function of a space-time rotation. These possibilities are:

$$d^2 = x^2 + y^2$$
$$d^2 = t^2 - z^2$$
$$d^2 = -x^2 - y^2$$
$$d^2 = -t^2 + z^2$$

(12.16)

In fact, there are only two possible forms here. One form has all the signs the same (either all pluses or all minuses); the other form has two different signs. Thus we need to consider only:

$$d^2 = t^2 + x^2 + y^2$$
$$d^2 = t^2 + x^2 - y^2$$

(12.17)

One form has all the signs the same (either all pluses or all minuses); the other form has two signs the same and one sign different.

In N-dimensional space, there are N variables. These N variables can be paired together to form 2-dimensional planes in:

$$(N-1)+(N-2)+...+2+1 = \frac{N(N-1)}{2}$$

(12.18)

ways[43].

---

[43] This is an arithmetic progression.

In 3-dimensions, the quadratic form distance function is of 2 possible forms. One form has all signs the same; the other form has one sign different:

$$d^2 = t^2 + x^2 + y^2$$
$$d^2 = t^2 + x^2 - y^2$$

(12.19)

In 3-dimensional space, there are three possible pairs of variables (2-dimensional planes).

If the 3-dimensional distance function is:

$$d^2 = t^2 + x^2 + y^2$$

(12.20)

then the three pairs of variables have the same sign, and so this space has three Euclidean rotational planes but no space-time rotational planes. If the 3-dimensional distance function is:

$$d^2 = t^2 + x^2 - y^2$$

(12.21)

then the only one pair of variables have the same sign and two pairs of variables have different signs, and so this space has one Euclidean rotational plane but two space-time rotational planes.

In 4-dimensional space, there are:

$$\frac{N(N-1)}{2} = \frac{4 \times 3}{2} = 6$$

(12.22)

2-dimensional planes. There are three possible different 4-dimensional distance functions:

$$d^2 = t^2 + x^2 + y^2 + z^2$$
$$d^2 = t^2 + x^2 + y^2 - z^2$$
$$d^2 = t^2 + x^2 - y^2 - z^2$$

(12.23)

(Note: $d^2 = t^2 - x^2 - y^2 - z^2$ is equivalent to $d^2 = t^2 + x^2 + y^2 - z^2$.)

In any dimension, the distance function with all the same signs will have:

$$\frac{N(N-1)}{2} \qquad (12.24)$$

Euclidean rotational planes. In any dimension, the distance function with only one different sign will have $(N-1)$ space-time rotational planes and thus:

$$\frac{N(N-1)}{2} - (N-1) = \frac{(N-1)(N-2)}{2} \qquad (12.25)$$

Euclidean rotational planes. In any dimension, the distance function with two different signs will have $2(N-2)$ space-time rotational planes and thus:

$$\frac{N(N-1)}{2} - 2(N-2) = \frac{N^2 - 5N + 8}{2} \qquad (12.26)$$

Euclidean rotational planes. Thus, a 4-dimensional emergent expectation space might be of the forms:

$$6 \quad \text{Euclidean rotations}$$
$$3 \quad \text{Euclidean rotations} \quad \& \quad 3 \quad \text{Space-time rotations}$$
$$2 \quad \text{Euclidean rotations} \quad \& \quad 4 \quad \text{Space-time rotations}$$
$$(12.27)$$

Clearly, our space-time is the 3 Euclidean rotation and 3 space-time rotations. If it were the case that the A_3 expectation distance function was of the form $d^2 = t^2 + x^2 - y^2 - z^2$, then we would have a space-time with 2 Euclidean rotations and 4 space-time rotations. The maths does not give this. However, the quaternion expectation distance function is of the form $d^2 = t^2 + x^2 + y^2 + z^2$, and so we must presume the existence of a quaternion emergent expectation space with 6 Euclidean rotations; there is no time in such a space, and it is not understood how such a timeless space manifests itself. We think the quaternion space is connected to the weak force.

In 5-dimensional space, there are $\dfrac{5(5-1)}{2}=10$ 2-dimensional planes. There are three possible different 5-dimensional distance functions:

$$d^2 = t^2 + w^2 + x^2 + y^2 + z^2$$
$$d^2 = t^2 + w^2 + x^2 + y^2 - z^2 \qquad (12.28)$$
$$d^2 = t^2 + w^2 + x^2 - y^2 - z^2$$

Thus, a 5-dimensional emergent expectation space might be of the forms:

10   Euclidean rotations

6   Euclidean rotations & 4   Space-time rotations

4   Euclidean rotations & 6   Space-time rotations

(12.29)

In 6-dimensional space, there are $\dfrac{6(6-1)}{2}=15$ 2-dimensional planes. There are four possible different 6-dimensional distance functions:

$$d^2 = t^2 + v^2 + w^2 + x^2 + y^2 + z^2$$
$$d^2 = t^2 + v^2 + w^2 + x^2 + y^2 - z^2$$
$$d^2 = t^2 + v^2 + w^2 + x^2 - y^2 - z^2 \qquad (12.30)$$
$$d^2 = t^2 + v^2 + w^2 - x^2 - y^2 - z^2$$

In any dimension, the distance function with three different signs will have $3(N-3)$ space-time rotational planes and thus:

$$\frac{N(N-1)}{2} - 3(N-3) = \frac{N^2 - 7N + 18}{2} \qquad (12.31)$$

Euclidean rotational planes. In general, in even dimensioned spaces with an equal number of minus signs and plus signs, there will be

$\dfrac{N^2}{4}$ space-time rotations. Thus, a 6-dimensional emergent expectation space might be of the forms:

15 Euclidean rotations

10 Euclidean rotations & 5 Space-time rotations

7 Euclidean rotations & 8 Space-time rotations

6 Euclidean rotations & 9 Space-time rotations

(12.32)

In 7-dimensional space, there are $\dfrac{7(7-1)}{2} = 21$ 2-dimensional planes. There are four possible different 7-dimensional distance functions:

$$d^2 = t^2 + u^2 + v^2 + w^2 + x^2 + y^2 + z^2$$
$$d^2 = t^2 + u^2 + v^2 + w^2 + x^2 + y^2 - z^2$$
$$d^2 = t^2 + u^2 + v^2 + w^2 + x^2 - y^2 - z^2 \qquad (12.33)$$

$$d^2 = t^2 + u^2 + v^2 + w^2 - x^2 - y^2 - z^2 \qquad (12.34)$$

Thus, a 7-dimensional emergent expectation space might be of the forms:

21 Euclidean rotations

15 Euclidean rotations & 6 Space-time rotations

11 Euclidean rotations & 10 Space-time rotations

9 Euclidean rotations & 12 Space-time rotations

(12.35)

Of course, we do not get 3-dimensional, 5-dimensional, 6-dimensional, or 7-dimensional spaces emerging from the $C_2 \times C_2 \times ...$ spinor algebras. We do get 8-dimensional emergent expectation spaces from the $C_2 \times C_2 \times C_2$ spinor algebras. These spaces are of great interest to us.

Fabrication of 8-dimensional emergent expectation spaces:

In 8-dimensional space, there are 28 2-dimensional planes. There are four possible different 8-dimensional distance functions:

$$d^2 = t^2 + s^2 + u^2 + v^2 + w^2 + x^2 + y^2 + z^2$$
$$d^2 = t^2 + s^2 + u^2 + v^2 + w^2 + x^2 + y^2 - z^2$$
$$d^2 = t^2 + s^2 + u^2 + v^2 + w^2 + x^2 - y^2 - z^2 \qquad (12.36)$$
$$d^2 = t^2 + s^2 + u^2 + v^2 + w^2 - x^2 - y^2 - z^2$$
$$d^2 = t^2 + s^2 + u^2 + v^2 - w^2 - x^2 - y^2 - z^2$$

Thus, a 8-dimensional emergent expectation space might be of the forms:

$$28 \quad \text{Euclidean rotations}$$
$$21 \quad \text{Euclidean rotations} \quad \& \quad 7 \quad \text{Space-time rotations}$$
$$16 \quad \text{Euclidean rotations} \quad \& \quad 12 \quad \text{Space-time rotations}$$
$$13 \quad \text{Euclidean rotations} \quad \& \quad 15 \quad \text{Space-time rotations}$$
$$(12.37)$$
$$12 \quad \text{Euclidean rotations} \quad \& \quad 16 \quad \text{Space-time rotations}$$
$$(12.38)$$

This is the result towards which we have been working. If any of the 8-dimensional $C_2 \times C_2 \times C_2$ spinor algebras produce an expectation distance function of the form of (12.36), we have an 8-dimensional emergent expectation space.

There are five distinct, two commutative and three non-commutative, 8-dimensional $C_2 \times C_2 \times C_2$ spinor algebras; there are 1024 8-dimensional algebras in total. We can search for 8-dimensional expectation distance functions by taking the 4th roots of the determinants of isomorphic spinor algebras, summing these rooted determinants, and factorising the expression[44]. Alternatively, more easily and more certainly, we can test for such a space by counting the number of pairs of variables which form either

---

[44] How does one prove that an expression of hundreds of terms cannot be factorised into the appropriate form?

Euclidean rotations or space-time rotations and compare these numbers to (12.38). If we have a match, we have an 8-dimensional emergent expectation space. If we do not have a match with (12.38), we do not have an 8-dimensional emergent expectation space.

If we do not have an 8-dimensional emergent expectation space, then we cannot have a 16-dimensional emergent expectation space from the $C_2 \times C_2 \times C_2 \times C_2$ spinor algebras because, if eight of the sixteen dimensions do not form a space, then the sixteen dimensions cannot form a space.

It is known that there are no 8-dimensional emergent expectation spaces, and so our 4-dimensional space-time is the only emergent expectation space to emerge from the $C_2 \times C_2 \times ...$ spinor algebras. Thus, we live in a unique 4-dimensional emergent expectation space. We'll say that again.

There is one and only one emergent expectation space which emerges from the $C_2 \times C_2 \times ...$ spinor algebras, and that emergent expectation space is our 4-dimensional space-time.

Emergent expectation space:
The expectation space emerging from a set of spinor algebras is formed by 'adding together' the isomorphic algebras in their different bases. This is just the same as taking the expectation value of electron spin in, say, the z-direction by adding the spins in the two possible different bases of the z-direction, up and down. We call this 'adding together' the isomorphic algebras the super-imposition of the algebras.

Any mathematician will tell you that you cannot add algebras that are written in different bases without destroying the algebra. They are all correct. So what are we left with when we have so destroyed the algebras? If we have destroyed the algebras, we have destroyed the algebraic multiplication operation, and so we cannot multiply. Since we cannot multiply, we cannot have imaginary variables like $i^2 = -1$. Nor can we have rotation because rotation is multiplication by a rotation matrix. Nor can we have commutation relations between variables. However, the variables still exist, but they must be real variables. We are left with only a continuous set of n-tuples of real numbers where n is the dimension of the isomorphic spinor algebras. We are left with a manifold which is the same thing as a set of co-ordinate points upon which we can place any co-ordinate system we choose.

Super-imposition of isomorphic spinor algebras produces a manifold of dimension equal to the dimension of each of the spinor algebras.

Expectation distance function:
Every spinor algebra has a norm (distance function). The distance function of a spinor algebra is the determinant of the matrix form of the algebra. The distance function of the hyperbolic complex numbers, (12.1), is:

$$\det\left(\begin{bmatrix} t & z \\ z & t \end{bmatrix}\right) = t^2 - z^2 \tag{12.39}$$

Putting this equal to the polar form of the algebra gives:

$$\det\left(\begin{bmatrix} t & z \\ z & t \end{bmatrix}\right) = \det\left(\begin{bmatrix} r & 0 \\ 0 & r \end{bmatrix}\begin{bmatrix} \cosh z & \sinh z \\ \sinh z & \cosh z \end{bmatrix}\right) \tag{12.40}$$
$$t^2 - z^2 = r^2$$

The distance functions of the six A_3 algebras are:

$$d^4 = \left(t^2 - x^2 - y^2 + z^2\right)^2$$
$$d^4 = \left(t^2 - x^2 - y^2 + z^2\right)^2$$
$$d^4 = \left(t^2 - x^2 + y^2 - z^2\right)^2$$
$$d^4 = \left(t^2 - x^2 + y^2 - z^2\right)^2 \tag{12.41}$$
$$d^4 = \left(t^2 + x^2 - y^2 - z^2\right)^2$$
$$d^4 = \left(t^2 + x^2 - y^2 - z^2\right)^2$$

We take the square roots of both sides to get the d^2 form. We then sum these d^2 forms to form the expectation distance function. Because we have taken the d^2 forms before summing, we get the d^2 form of the expectation distance function. This is the form that is conserved by the 2-dimensional spinor rotations (Euclidean and space-time (Lorentz boost) rotations), and so this will allow fabrication of an emergent expectation space from the 2-dimensional spinor spaces. Such a fabricated space has geometry described by Riemann mathematics (tensor calculus etc.).

$$SUM \begin{cases} d^2 = t^2 - x^2 - y^2 + z^2 \\ d^2 = t^2 - x^2 - y^2 + z^2 \\ d^2 = t^2 - x^2 + y^2 - z^2 \\ d^2 = t^2 - x^2 + y^2 - z^2 \\ d^2 = t^2 + x^2 - y^2 - z^2 \\ d^2 = t^2 + x^2 - y^2 - z^2 \end{cases} = 2\left(3t^2 - x^2 - y^2 - z^2\right) \quad (12.42)$$

Wherein we have taken the expectation distance function by simply adding the individual distance functions. We have the distance function of our space-time as the expectation distance function of the A_3 spinor algebras.

Within traditional Riemann geometry, it is assumed that there is a 3-dimensional space with distance function:

$$d^2 = x^2 + y^2 + z^2 \quad (12.43)$$

The only 3-dimensional distance function that emerges from the 3-dimensional spinor algebras is:

$$SUM \begin{cases} a^3 - b^3 + c^3 + 3abc \\ a^3 + b^3 - c^3 + 3abc \end{cases} = 2a^3 + 6abc \quad (12.44)$$

If we are to derive our distance functions from super-imposition of algebraically isomorphic spinor algebras, we are going to have a very much restricted list of different distance functions.

Angles in emergent expectation spaces:
When we superimpose isomorphic algebras written in different bases, we destroy the algebraic structure of those algebras. Part of that algebraic structure is rotation (polar form of the algebra) and the type of angles in that rotation. A 4-dimensional quaternion angle is a very different thing from a 2-dimensional hyperbolic angle. Without rotation, the manifold which emerges from super-imposition has no concept of rotation or of angles within it; without rotation and angles, there no concept of curvature.

When we superimpose the 2-dimensional spinor algebras to get the emergent expectation space, because there is only one copy of each algebra (it is written in only one basis), the algebraic structure survives super-imposition. That algebraic structure includes 2-dimensional rotation and the associated angles.

The distance functions of the two 2-dimensional spinor algebras are preserved by the 2-dimensional rotations. The 2-dimensional distance functions are sub-functions of the 4-dimensional expectation distance function which emerges from the super-imposition of the A_3 distance functions. It is thus possible for the 2-dimensional rotations to exist within our 4-dimensional space-time. Indeed, it seems they form a large part of the structure of our 4-dimensional space-time. The metric and the entire notion of curvature comes to mind.

Within the 2-dimensional spinor algebras, there are 2-dimensional inner products and 2-dimensional curls. Since our 4-dimensional space-time can support 2-dimensional rotations, it can also support the other aspects of the 2-dimensional spinor algebras. That is why we have 2-dimensional inner products and 2-dimensional curls in our 4-dimensional space-time. We see the 2-dimensional curls in many classical physics equations.

Rotations in other emergent expectation spaces:
The reader might wonder if there are any other expectation distance functions from the super-imposition of other spinor algebras, perhaps based on other finite groups, the group C_4 perhaps, which are such that they contain as sub-functions the 2-dimensional

distance functions and thus allow the 2-dimensional rotations to exist within them. This is not yet clearly understood. The sub-group structure of the group is involved, but this is not a simple involvement. The two emergent expectation spaces of the group C_4 might allow a 2-dimensional rotation in a single 2-dimensional plane, but not three 2-dimensional rotations in three orthogonal planes; even this seems unlikely. A more interesting candidate is the group $C_2 \times C_2 \times C_2$ which is seemingly closely connected to the strong force. However, the 8-dimensional spinor spaces in the $C_2 \times C_2 \times C_2$ algebras 'fold up' into 4-dimensional spinor spaces, and so there are unanswered questions regarding the nature of the 8-dimensional spaces.

There are two 3-dimensional spinor algebras that come through the super-imposition unscathed. Are there emergent expectation spaces, perhaps from the group S_3, that allow 3-dimensional rotations? It seems not, but this is not thoroughly understood.

It seems, and this is not proven or properly understood, that the only emergent expectation space which allows rotation in every pair (or trio in the 3-dimensional angle case) of axes is our 4-dimensional space-time. There's a thing for the reader to cogitate upon.

More consequences of 2-dimensional angles:
Curvature in our 4-dimensional space-time must be curvature of a 2-dimensional nature. What else can it be, we have only 2-dimensional rotations within our space-time. Well, it might be 4-dimensional curvature associated with the underlying A_3 algebras. There are similarities between the 4-dimensional 2-dimensional rotations of the A_3 algebras and the 2-dimensional rotations. The 4-dimensional 2-dimensional rotations are double covers of the 2-dimensional rotations. However, if it were 4-dimensional curvature, we would have only one 4-dimensional principle curvature; we have six principle curvatures.

The 3-dimensional spinor algebras, from the group C_3, have within them 3-dimensional rotation and the associated 3-dimensional angles. Such a rotation is:

$$\begin{bmatrix} v_A(\theta,\phi) & v_B(\theta,\phi) & v_C(\theta,\phi) \\ v_C(\theta,\phi) & v_A(\theta,\phi) & v_B(\theta,\phi) \\ v_B(\theta,\phi) & v_C(\theta,\phi) & v_A(\theta,\phi) \end{bmatrix} \qquad (12.45)$$

where v_i are the nu-functions which are the trigonometric functions of this 3-dimensional spinor space.[45] Try to imagine curvature based on this 3-dimensional rotation; it would be very different to the curvature, based on our two types of 2-dimensional rotation, to which we are accustomed.

The Riemann curvature tensor is derived from the metric tensor. The components of the metric tensor are the inner products of two vectors. These inner products are based on the 2-dimensional inner products we find in the 2-dimensional spinor spaces. We therefore take the view that the curvature of space-time is of a 2-dimensional nature rather than of a 4-dimensional, double cover, nature. We think, that, if the curvature of space-time was 4-dimensional, we would see some effect of the doubly cover nature of that curvature. We do not see any such effects.

The metric tensor and the affine connection:
The form of the components of the metric tensor are effectively the 2-dimensional inner products that exist in only the 2-dimensional spinor algebras. The angles measured by these inner products are 2-dimensional angles. The metric varies from point to point over the 4-dimensional A_3 emergent expectation manifold. This varying is variation in the orientations of the 2-dimensional planes towards each other and the adjustments to length associated with these changes of orientations. We call this curvature. However, this makes sense only if we have an underlying affine connection that determines what is meant by orientation. We think that affine

---

[45] They were called nu-functions because they were new when first discovered.

structure is determined by the locally varying expectation A_3 phase (angle). Locally varying means the angle varies from point to point in the manifold. An example of an A_3 rotation matrix is:

$$A_{3\,Rot} = \begin{bmatrix} \cosh(\lambda) & \frac{b}{\lambda}\sinh(\lambda) & \frac{c}{\lambda}\sinh(\lambda) & \frac{d}{\lambda}\sinh(\lambda) \\ -\frac{b}{\lambda}\sinh(\lambda) & \cosh(\lambda) & -\frac{d}{\lambda}\sinh(\lambda) & \frac{c}{\lambda}\sinh(\lambda) \\ \frac{c}{\lambda}\sinh(\lambda) & -\frac{d}{\lambda}\sinh(\lambda) & \cosh(\lambda) & -\frac{b}{\lambda}\sinh(\lambda) \\ \frac{d}{\lambda}\sinh(\lambda) & \frac{c}{\lambda}\sinh(\lambda) & \frac{b}{\lambda}\sinh(\lambda) & \cosh(\lambda) \end{bmatrix}$$

$$\lambda = \sqrt{-b^2 + c^2 + d^2}$$

(12.46)

This is only a few minus signs different from the quaternion rotation matrix, (12.10). Setting any two of the variables in this to zero will produce 4-dimensional 2-dimensional rotations - don't remove the square root signs.

Within QFT, the photon field emerges by allowing local variation of the orientation of a vector (an angle) in the complex plane, $\mathbb{C}$, over a flat underlying space. Such local variation makes sense only if the underlying space has within it an affine connection (a sense of what is meant by parallel) against which the orientation of the varying vector can be measured. In short, the underlying space has an infinitude of copies of the complex plane fixed to it, one at each point pinned through the origin; it is a fibre bundle. Within QFT, the weak force emerges by allowing local variation of the orientation of a vector in $SU(2)$ space. We take $SU(2)$ space to be quaternion space. Within QFT, the strong force emerges by allowing local variation of the orientation of a vector in $SU(3)$ space[46].

---

[46] Your author opines that the $SU(3)$ bit is wrong and we ought to use the 8-dimensional Clifford algebra instead. This give six gluons not eight.

We think the gravitational force, curvature of space-time, emerges by allowing local variation of the orientation of a vector in A_3 space. You see the pattern.

Curvature:
But wait, just as the local variation of phase so prominent in QFT is meaningless without an affine connection in the underlying space, so local variation of the orientation of a vector in A_3 space over a manifold with no affine connection is meaningless. What would local variation of a phase over a manifold with no affine connection look like? We think the local variation of the orientation of a vector in A_3 space over the emergent expectation manifold induces the affine connection into the manifold. This is why gravity is different from the other forces of nature. This is why the gravitational field is within space-time while other force fields are over space-time. We are asserting that the affine connection of our 4-dimensional space-time is local variation of A_3 phase.

Since the group $C_2 \times C_2$ has three C_2 sub-groups, A_3 rotation has three 2-dimensional rotational planes within it. These are 4-dimensional 2-dimensional planes, but the structure is such that it will allow 2-dimensional rotations to 'fit together' orthogonally. Local variation of the expectation A_3 phase space over the emergent manifold will distort this 'fitting together' of the 2-dimensional rotation planes. This distortion is curvature measured by the, in 4-dimensional space, twenty independent components of the Riemann curvature tensor.

Local flatness:
We have two reasons to expect local flatness in our 4-dimensional space-time.

1) Every spinor algebra is flat. The emergent expectation space that arises from the super-imposition of six such flat algebras will be flat at an infinitesimally small point. Outside of the point, the algebras will fall to bits and flatness is meaningless.

2) The metric tensor is formed from inner products based on 2-dimensional inner products. All inner products of spinor algebras are symmetric, $\vec{e_1} \cdot \vec{e_2} = \vec{e_2} \cdot \vec{e_1}$, and so the metric is symmetric. It is a mathematical fact that a symmetric matrix can be made diagonal. The metric tensor varies from point to point in our space-time, and so the diagonal of the matrix will vary from point to point over space-time. We associate a diagonal matrix with local flatness. Thus we see that local flatness is consistent with the adoption of the 2-dimensional inner products in forming the 4-dimensional inner products that form the metric tensor of our space-time.

In fact these two reasons are tied together within a spinor algebra. Because real numbers are commutative and the inner product in any spinor algebra is a real number, we must have symmetry. If the spinor algebra was not flat, the algebra would fall to bits and we would have no inner product.

Other spinor algebras:

There are two 2-dimensional spinor algebras within the group C_2. Within the group C_3, there are four 3-dimensional spinor algebras which do not interact with our space-time because they are commutative. Within the group C_4 there are eight 4-dimensional commutative spinor algebras which do not interact with our space-time, and within the group $C_2 \times C_2$, there are eight 4-dimensional commutative spinor algebras which do not interact with our space-time. Also within $C_2 \times C_2$ there are the quaternions, the anti-quaternions, and the six A_3 algebras. We have the photon field associated with the 2-dimensional complex numbers, $\mathbb{C}$. We have the weak force associated with the quaternions. We can associate an anti-weak force with the anti-quaternions which matches the weak force and would have no observational consequences. We are left with the 2-dimensional hyperbolic complex numbers, $\mathbb{S}$; where do they fit in?

Suppose we allow, over an underlying space, local variation of the orientation, phase, of a vector in the 2-dimensional (special relativity) space-time that is the hyperbolic complex numbers. In this algebra, change of angle is change of velocity; local change of angle is local change of velocity; an object accelerates in moving from one point to another. Change of velocity is associated with a force and a proportionality constant called inertial mass. We see a local change of phase is associated with a force. It is a force that is neither gravitational, electromagnetic, weak nor strong. It is an often ignored force, but it really is a force as anyone who has been hit by a cannon ball will attest.

Wot! no gravitons:

Within QFT, there are bosons associated with the locally varying orientation of the vectors. The photon is associated with the local variation of the orientation of a vector (an angle) in the complex plane, $\mathbb{C}$. Why are there no bosons associated with the local variation of the orientation of a vector in A_3 space? We think the answer is because there is no established affine connection in the emergent expectation manifold. We predict the non-existence of gravitons.

Summary:

We have laid out above how we think our 4-dimensional curved space-time with its rotations, angles, distance function, metric, affine connection and local flatness emerges from the spinor algebras. We do not yet have an energy-momentum tensor. We rectify that in the next chapter.

Chapter 13

The Energy-Momentum Tensor and Electromagnetism

We have shown elsewhere[47] how to differentiate a non-commutative spinor potential (a Clifford algebra if you prefer) to get the $E \& B$ fields:

$$E = \frac{1}{2}(d_L + d_R)$$
$$B = \frac{1}{2}(d_L - d_R)$$

(13.1)

We begin with a potential of each of the six A_3 algebras and calculate the E field and the B field by differentiation of the potential each algebra. An example of an A_3 potential is:

$$SSA_{Pot} = \begin{bmatrix} \phi & A_x & A_y & A_z \\ A_x & \phi & A_z & A_y \\ A_y & -A_z & \phi & -A_x \\ -A_z & A_y & -A_x & \phi \end{bmatrix}$$

(13.2)

Conventionally, we would see ϕ as a scalar potential and A_i as a vector potential. Having calculated the six E fields and the six B fields. We then take the expectation field (expectation tensor) by adding the twelve fields together. This produces the expectation field tensor:

$$Div = \left(\frac{\partial \phi}{\partial t} + \frac{\partial A_x}{\partial x} + \frac{\partial A_y}{\partial y} + \frac{\partial A_z}{\partial z} \right)$$

(13.3)

---

[47] See Dennis Morris The Physics of Empty Space.

$$4\begin{bmatrix} 3\,Div & 3\dfrac{\partial A_x}{\partial t}+\dfrac{\partial \phi}{\partial x} & 3\dfrac{\partial A_y}{\partial t}+\dfrac{\partial \phi}{\partial y} & 3\dfrac{\partial A_z}{\partial t}+\dfrac{\partial \phi}{\partial z} \\[2ex] \dfrac{\partial A_x}{\partial t}+3\dfrac{\partial \phi}{\partial x} & 3\,Div & 3\dfrac{\partial A_y}{\partial x}+\dfrac{\partial A_x}{\partial y} & 3\dfrac{\partial A_z}{\partial x}+\dfrac{\partial A_x}{\partial z} \\[2ex] \dfrac{\partial A_y}{\partial t}+3\dfrac{\partial \phi}{\partial y} & \dfrac{\partial A_y}{\partial x}+3\dfrac{\partial A_x}{\partial y} & 3\,Div & 3\dfrac{\partial A_z}{\partial y}+\dfrac{\partial A_y}{\partial z} \\[2ex] \dfrac{\partial A_z}{\partial t}+3\dfrac{\partial \phi}{\partial z} & \dfrac{\partial A_z}{\partial x}+3\dfrac{\partial A_x}{\partial z} & \dfrac{\partial A_z}{\partial y}+3\dfrac{\partial A_y}{\partial z} & 3\,Div \end{bmatrix} \quad (13.4)$$

From this tensor, we extract the symmetric part:

$$8\begin{bmatrix} Div & \dfrac{\partial A_x}{\partial t}+\dfrac{\partial \phi}{\partial x} & \dfrac{\partial A_y}{\partial t}+\dfrac{\partial \phi}{\partial y} & \dfrac{\partial A_z}{\partial t}+\dfrac{\partial \phi}{\partial z} \\[2ex] \dfrac{\partial A_x}{\partial t}+\dfrac{\partial \phi}{\partial x} & Div & \dfrac{\partial A_y}{\partial x}+\dfrac{\partial A_x}{\partial y} & \dfrac{\partial A_z}{\partial x}+\dfrac{\partial A_x}{\partial z} \\[2ex] \dfrac{\partial A_y}{\partial t}+\dfrac{\partial \phi}{\partial y} & \dfrac{\partial A_y}{\partial x}+\dfrac{\partial A_x}{\partial y} & Div & \dfrac{\partial A_z}{\partial y}+\dfrac{\partial A_y}{\partial z} \\[2ex] \dfrac{\partial A_z}{\partial t}+\dfrac{\partial \phi}{\partial z} & \dfrac{\partial A_z}{\partial x}+\dfrac{\partial A_x}{\partial z} & \dfrac{\partial A_z}{\partial y}+\dfrac{\partial A_y}{\partial z} & Div \end{bmatrix} \quad (13.5)$$

And the anti-symmetric part:

$$4\begin{bmatrix} Div & \dfrac{\partial A_x}{\partial t}-\dfrac{\partial \phi}{\partial x} & \dfrac{\partial A_y}{\partial t}-\dfrac{\partial \phi}{\partial y} & \dfrac{\partial A_z}{\partial t}-\dfrac{\partial \phi}{\partial z} \\[2ex] -\left(\dfrac{\partial A_x}{\partial t}-\dfrac{\partial \phi}{\partial x}\right) & Div & \dfrac{\partial A_y}{\partial x}-\dfrac{\partial A_x}{\partial y} & \dfrac{\partial A_x}{\partial z}-\dfrac{\partial A_z}{\partial x} \\[2ex] -\left(\dfrac{\partial A_y}{\partial t}-\dfrac{\partial \phi}{\partial y}\right) & -\left(\dfrac{\partial A_y}{\partial x}-\dfrac{\partial A_x}{\partial y}\right) & Div & \dfrac{\partial A_z}{\partial y}-\dfrac{\partial A_y}{\partial z} \\[2ex] -\left(\dfrac{\partial A_z}{\partial t}-\dfrac{\partial \phi}{\partial z}\right) & -\left(\dfrac{\partial A_z}{\partial x}-\dfrac{\partial A_x}{\partial z}\right) & -\left(\dfrac{\partial A_z}{\partial y}-\dfrac{\partial A_y}{\partial z}\right) & Div \end{bmatrix}$$

$$(13.6)$$

Clearly, we have (13.5) + (13.6) = (13.4).

We notice that the elements of the symmetric tensor are of the form of the 2-dimensional space-time curl which we find in the hyperbolic complex numbers, $\mathbb{S}$. We assert that the divergence is zero in both the symmetric and anti-symmetric parts. We certainly need it to be zero in the anti-symmetric part, (13.6). To have left all twelve copies of the divergence within the symmetric part would have spoilt the balance of the symmetric part.

The energy-momentum tensor:
We have a symmetric tensor emerging as the symmetric part of the A_3 emergent expectation tensor; we assume that this is the energy-momentum tensor, $T^{\mu\nu}$.

The electromagnetic tensor:
We are convinced that the anti-symmetric part of the A_3 emergent expectation tensor is the electromagnetic field tensor. We have elsewhere dealt with electromagnetism emerging from the quaternion part of the $C_2 \times C_2$ group. The anti-symmetric variables of the six A_3 algebras taken together form the quaternion and the anti-quaternion algebras. Within the anti-symmetric part (13.6), we put:

$$E_t = Div \qquad\qquad B_t = 0$$

$$E_x = \frac{\partial A_x}{\partial t} - \frac{\partial \phi}{\partial x} \qquad B_x = \frac{\partial A_z}{\partial y} - \frac{\partial A_y}{\partial z}$$

$$E_y = \frac{\partial A_y}{\partial t} - \frac{\partial \phi}{\partial y} \qquad B_y = \frac{\partial A_x}{\partial z} - \frac{\partial A_z}{\partial x} \qquad (13.7)$$

$$E_z = \frac{\partial A_z}{\partial t} - \frac{\partial \phi}{\partial z} \qquad B_z = \frac{\partial A_y}{\partial x} - \frac{\partial A_x}{\partial y}$$

These definitions do not quite match the conventional definitions of the electric and magnetic fields. Which are:

$$E_t = Div \qquad\qquad B_t = 0$$

$$E_x = -\frac{\partial A_x}{\partial t} - \frac{\partial \phi}{\partial x} \qquad B_x = \frac{\partial A_z}{\partial y} - \frac{\partial A_y}{\partial z}$$

$$E_y = -\frac{\partial A_y}{\partial t} - \frac{\partial \phi}{\partial y} \qquad B_y = \frac{\partial A_x}{\partial z} - \frac{\partial A_z}{\partial x} \qquad (13.8)$$

$$E_z = -\frac{\partial A_z}{\partial t} - \frac{\partial \phi}{\partial z} \qquad B_z = \frac{\partial A_y}{\partial x} - \frac{\partial A_x}{\partial y}$$

The difference between (13.8) and (13.7) is merely the conventional arbitrary definition of the electric field. By reversing the defined direction of the electric field (so that current runs from negative to positive as we now know it does) and changing the defined sign of the scalar potential, we get a perfect match. We notice in (13.7) that the electric field comes as a curl. We usually associate forces with curls, and so perhaps this is a better definition of the electric field than the conventional one. It seems that the mathematics has given electromagnetism as it ought to be defined.

This enables us to rewrite the anti-symmetric part of the emergent A_3 tensor as:

$$F^{\mu\nu} = \begin{bmatrix} 0 & E_x & E_y & E_z \\ -E_x & 0 & B_z & B_y \\ -E_y & -B_z & 0 & B_x \\ -E_z & -B_y & -B_x & 0 \end{bmatrix} \qquad (13.9)$$

We compare this to the conventional electromagnetic tensor which is[48]:

$$F^{\mu\nu} = \begin{bmatrix} 0 & -E_x & -E_y & -E_z \\ E_x & 0 & -B_z & B_y \\ E_y & B_z & 0 & -B_x \\ E_z & -B_y & B_x & 0 \end{bmatrix} \qquad (13.10)$$

---

[48] See: David McMahon Quantum Field Theory Demystified, Pg 165.

We see that reversing the axes $\{x, y, z\}$ and reversing the conventionally defined direction of the $Curl_y$ field, it always was the odd one out, we get the conventional electromagnetic field tensor.

We note that we have previously derived electromagnetism from only the quaternion algebras. We now opine that, although deriving electromagnetism from the quaternion algebras is very instructive, the A_3 algebras are the source of electromagnetism. We opine that the quaternion algebras are the source of the weak force.

Three generations of matter:
Physicists have been puzzled by the observation that particles come in three generations. For example, we have the electron, the muon and the tau particle. We also have three generations of quarks, up & down, strange & charm, top and bottom. The three particles in each set are identical except for mass. It is as if each particle has three distinct masses, and we prefer this view to the three generations view.

There are three pairs of A_3 algebras. If, as it seems, the charge of the emergent symmetric A_3 tensor is mass, then we might expect each particle would have a mass corresponding to the symmetric part of a pair of A_3 algebras. Hence, we predict three generations of matter. This fits with observation.

The Maxwell type equations:
Within the non-commutative $C_2 \times C_2$ spinor algebras (these are the quaternions and the A_3 algebras and their anti-algebras), it is a standard differential identity[49] that:

$$\{B, d\} = [E, d] \qquad (13.11)$$

---

[49] See Dennis Morris: The Physics of Empty Space

where curly-B and straight-E are second differentials of the potential and first differentials of the $B \& E$ fields of that potential.

The electromagnetic homogeneous Maxwell equations:
We take the expectation curly-B field by adding all the A_3 curly-B fields, and we take the expectation straight-E field by adding all the A_3 straight-E fields. This gives:

$$\sum_{\text{All algebras}} [E,d] = 4 \begin{bmatrix} 0 & 0 & 0 & 0 \\ 0 & 0 & 3\dfrac{\partial E_y}{\partial x}+\dfrac{\partial E_x}{\partial y} & 3\dfrac{\partial E_z}{\partial x}+\dfrac{\partial E_x}{\partial z} \\ 0 & \dfrac{\partial E_y}{\partial x}+3\dfrac{\partial E_x}{\partial y} & 0 & 3\dfrac{\partial E_z}{\partial y}+\dfrac{\partial E_y}{\partial z} \\ 0 & \dfrac{\partial E_z}{\partial x}+3\dfrac{\partial E_x}{\partial z} & \dfrac{\partial E_z}{\partial y}+3\dfrac{\partial E_y}{\partial z} & 0 \end{bmatrix}$$

(13.12)

$$\sum_{\text{All algebras}} \{B,d\} = 4 \begin{bmatrix} 3Div & 3\dfrac{\partial B_x}{\partial t} & 3\dfrac{\partial B_y}{\partial t} & 3\dfrac{\partial B_z}{\partial t} \\ \dfrac{\partial B_x}{\partial t} & 3Div & 0 & 0 \\ \dfrac{\partial B_y}{\partial t} & 0 & 3Div & 0 \\ \dfrac{\partial B_z}{\partial t} & 0 & 0 & 3Div \end{bmatrix}$$

(13.13)

$$Div = \frac{\partial B_x}{\partial x} + \frac{\partial B_y}{\partial y} + \frac{\partial B_z}{\partial z} \qquad (13.14)$$

We assume that, since, within any A_3 algebra, curly-B is equal to straight-E, the expectation curly-B is equal to the expectation straight-E. We will in due course extract the Maxwell type equations by putting these two matrices, (13.12) & (13.13) equal. Before we do that, we separate out the symmetric and anti-symmetric parts of each matrix:

The straight-E fields give:

$$\sum_{\substack{\text{All algebras}}} \left[E,d\right]_{Sym} = 8 \begin{bmatrix} 0 & 0 & 0 & 0 \\ 0 & 0 & \dfrac{\partial E_y}{\partial x}+\dfrac{\partial E_x}{\partial y} & \dfrac{\partial E_z}{\partial x}+\dfrac{\partial E_x}{\partial z} \\ 0 & \dfrac{\partial E_x}{\partial y}+\dfrac{\partial E_x}{\partial y} & 0 & \dfrac{\partial E_z}{\partial y}+\dfrac{\partial E_y}{\partial z} \\ 0 & \dfrac{\partial E_z}{\partial x}+\dfrac{\partial E_x}{\partial z} & \dfrac{\partial E_y}{\partial z}+\dfrac{\partial E_z}{\partial y} & 0 \end{bmatrix}$$

(13.15)

And:

$$\sum_{\substack{\text{All algebras}}} \left[E,d\right]_{Anti} = 4 \begin{bmatrix} 0 & 0 & 0 & 0 \\ 0 & 0 & \dfrac{\partial E_y}{\partial x}-\dfrac{\partial E_x}{\partial y} & \dfrac{\partial E_z}{\partial x}-\dfrac{\partial E_x}{\partial z} \\ 0 & -\left(\dfrac{\partial E_y}{\partial x}-\dfrac{\partial E_x}{\partial y}\right) & 0 & \dfrac{\partial E_z}{\partial y}-\dfrac{\partial E_y}{\partial z} \\ 0 & -\left(\dfrac{\partial E_z}{\partial x}-\dfrac{\partial E_x}{\partial z}\right) & -\left(\dfrac{\partial E_z}{\partial y}-\dfrac{\partial E_y}{\partial z}\right) & 0 \end{bmatrix}$$

(13.16)

The curly-B fields give:

$$\sum_{\substack{\text{All algebras}}} \left\{B,d\right\}_{Sym} = 8 \begin{bmatrix} Div & \dfrac{\partial B_x}{\partial t} & \dfrac{\partial B_y}{\partial t} & \dfrac{\partial B_z}{\partial t} \\ \dfrac{\partial B_x}{\partial t} & Div & 0 & 0 \\ \dfrac{\partial B_y}{\partial t} & 0 & Div & 0 \\ \dfrac{\partial B_z}{\partial t} & 0 & 0 & Div \end{bmatrix}$$

(13.17)

And:

$$\sum_{\substack{\text{All algebras}}} \{B,d\}_{Anti} = 4 \begin{bmatrix} Div & \dfrac{\partial B_x}{\partial t} & \dfrac{\partial B_y}{\partial t} & \dfrac{\partial B_z}{\partial t} \\[2ex] -\dfrac{\partial B_x}{\partial t} & Div & 0 & 0 \\[2ex] -\dfrac{\partial B_y}{\partial t} & 0 & Div & 0 \\[2ex] -\dfrac{\partial B_z}{\partial t} & 0 & 0 & Div \end{bmatrix} \qquad (13.18)$$

We have shown previously that electromagnetism is of a quaternion form[50]. Within each A_3 algebra, there is a single anti-symmetric variable. Three of these variables taken together form a quaternion; the other three taken together form an anti-quaternion. The three A_3 algebras which contain quaternion type anti-symmetric variables are $\{SSA^*, SAS, ASS\}$ [51]. We extract the anti-symmetric variables from these three algebras and require the sum of the two anti-symmetric matrices to be of quaternion form. Summing the two anti-symmetric matrices gives:

$$4 \begin{bmatrix} Div & \dfrac{\partial B_x}{\partial t} & \dfrac{\partial B_y}{\partial t} & \dfrac{\partial B_z}{\partial t} \\[2ex] -\dfrac{\partial B_x}{\partial t} & Div & \dfrac{\partial E_y}{\partial x} - \dfrac{\partial E_x}{\partial y} & \dfrac{\partial E_z}{\partial x} - \dfrac{\partial E_x}{\partial z} \\[2ex] -\dfrac{\partial B_y}{\partial t} & -\left(\dfrac{\partial E_y}{\partial x} - \dfrac{\partial E_x}{\partial y}\right) & Div & \dfrac{\partial E_z}{\partial y} - \dfrac{\partial E_y}{\partial z} \\[2ex] -\dfrac{\partial B_z}{\partial t} & -\left(\dfrac{\partial E_z}{\partial x} - \dfrac{\partial E_x}{\partial z}\right) & -\left(\dfrac{\partial E_z}{\partial y} - \dfrac{\partial E_y}{\partial z}\right) & Div \end{bmatrix}$$

$$(13.19)$$

For this matrix to be a quaternion form (see (12.4)) requires:

---

[50] Dennis Morris: The Physics of Empty Space
[51] It seems that we originally misnamed the *SSA* algebra.

$$\frac{\partial B_x}{\partial t} = \frac{\partial E_y}{\partial z} - \frac{\partial E_z}{\partial y}$$

$$\frac{\partial B_y}{\partial t} = \frac{\partial E_z}{\partial x} - \frac{\partial E_x}{\partial z}$$

$$\frac{\partial B_z}{\partial t} = \frac{\partial E_x}{\partial y} - \frac{\partial E_y}{\partial x}$$

(13.20)

These are the electromagnetic homogeneous Maxwell equations. As shown elsewhere, the non-appearance of the anti-quaternions is why we live in a classical matter universe without anti-matter.

The GEM homogeneous Maxwell equations:

We do with the symmetric parts of the emergent A_3 tensor the same as we did with the anti-symmetric parts. Summing the two symmetric matrices gives:

$$\sum_{\text{All algebras}} \{B,d\}_{Sym} + \sum_{\text{All algebras}} [E,d]_{Sym} =$$

$$8 \begin{bmatrix} Div & \dfrac{\partial B_x}{\partial t} & \dfrac{\partial B_y}{\partial t} & \dfrac{\partial B_z}{\partial t} \\[2mm] \dfrac{\partial B_x}{\partial t} & Div & \dfrac{\partial E_y}{\partial x}+\dfrac{\partial E_x}{\partial y} & \dfrac{\partial E_z}{\partial x}+\dfrac{\partial E_x}{\partial z} \\[2mm] \dfrac{\partial B_y}{\partial t} & \dfrac{\partial E_x}{\partial y}+\dfrac{\partial E_x}{\partial y} & Div & \dfrac{\partial E_z}{\partial y}+\dfrac{\partial E_y}{\partial z} \\[2mm] \dfrac{\partial B_z}{\partial t} & \dfrac{\partial E_z}{\partial x}+\dfrac{\partial E_x}{\partial z} & \dfrac{\partial E_y}{\partial z}+\dfrac{\partial E_z}{\partial y} & Div \end{bmatrix}$$

(13.21)

What is this object? If it is to be of the same form as an A_3 algebra, which of the A_3 algebras should we choose? We choose to make it equal to the symmetric parts of the same three A_3 algebras we used for the anti-symmetric matrices. These are the $\{SSA^*, SAS, ASS\}$ algebras. There are two symmetric variables in each of these three A_3 algebras. Requiring the matrix (13.21) to be of the form of the symmetric variables of these three A_3 algebras leads to the six equations:

$$\frac{\partial B_x}{\partial t} = +\left(\frac{\partial E_z}{\partial y} + \frac{\partial E_y}{\partial z}\right) \quad \frac{\partial B_x}{\partial t} = -\left(\frac{\partial E_z}{\partial y} + \frac{\partial E_y}{\partial z}\right)$$

$$\frac{\partial B_y}{\partial t} = +\left(\frac{\partial E_z}{\partial x} + \frac{\partial E_x}{\partial z}\right) \quad \frac{\partial B_y}{\partial t} = -\left(\frac{\partial E_z}{\partial x} + \frac{\partial E_x}{\partial z}\right) \qquad (13.22)$$

$$\frac{\partial B_z}{\partial t} = +\left(\frac{\partial E_y}{\partial x} + \frac{\partial E_x}{\partial y}\right) \quad \frac{\partial B_z}{\partial t} = -\left(\frac{\partial E_y}{\partial x} + \frac{\partial E_x}{\partial y}\right)$$

Clearly:

$$\frac{\partial B_x}{\partial t} = 0 \quad \left(\frac{\partial E_z}{\partial y} + \frac{\partial E_y}{\partial z}\right) = 0$$

$$\frac{\partial B_y}{\partial t} = 0 \quad \left(\frac{\partial E_z}{\partial x} + \frac{\partial E_x}{\partial z}\right) = 0 \qquad (13.23)$$

$$\frac{\partial B_z}{\partial t} = 0 \quad \left(\frac{\partial E_y}{\partial x} + \frac{\partial E_x}{\partial y}\right) = 0$$

Within each separate A_3 algebra, the Maxwell equations of that algebra are not zero. Within each A_3 algebra we have gravito-electromagnetism; there are a few signs different from conventional gravito-electromagnetism. Taken as a whole, in the classical universe, the gravito-electromagnetic force reduces to zero.[52]

The inhomogeneous Maxwell equations:
Above, we summed the six curly-B and the six straight-E fields. We now sum the curly-E fields and the straight-B fields.

---

[52] We have assumed the three A_3 algebras we used are in balance. If they are out of balance at some location in the universe, we would have a gravito-electromagnetic force at that location.

$$\sum_{\text{All algebras}} [B,d] = 4 \begin{bmatrix} 0 & 0 & 0 & 0 \\ 0 & 0 & 3\dfrac{\partial B_y}{\partial x}+\dfrac{\partial B_x}{\partial y} & 3\dfrac{\partial B_z}{\partial x}+\dfrac{\partial B_x}{\partial z} \\ 0 & \dfrac{\partial B_y}{\partial x}+3\dfrac{\partial B_x}{\partial y} & 0 & 3\dfrac{\partial B_z}{\partial y}+\dfrac{\partial B_y}{\partial z} \\ 0 & \dfrac{\partial B_z}{\partial x}+3\dfrac{\partial B_x}{\partial z} & \dfrac{\partial B_z}{\partial y}+3\dfrac{\partial B_y}{\partial z} & 0 \end{bmatrix}$$

$$(13.24)$$

$$\sum_{\text{All algebras}} \{E,d\} =$$

$$4 \begin{bmatrix} 3Div & 3\dfrac{\partial E_x}{\partial t}+\dfrac{\partial E_t}{\partial x} & 3\dfrac{\partial E_y}{\partial t}+\dfrac{\partial E_t}{\partial y} & 3\dfrac{\partial E_z}{\partial t}+\dfrac{\partial E_t}{\partial z} \\ \dfrac{\partial E_x}{\partial t}+3\dfrac{\partial E_t}{\partial x} & 3Div & 0 & 0 \\ \dfrac{\partial E_y}{\partial t}+3\dfrac{\partial E_t}{\partial y} & 0 & 3Div & 0 \\ \dfrac{\partial E_z}{\partial t}+3\dfrac{\partial E_t}{\partial z} & 0 & 0 & 3Div \end{bmatrix}$$

$$(13.25)$$

$$Div = \frac{\partial E_t}{\partial t}+\frac{\partial E_x}{\partial x}+\frac{\partial E_y}{\partial y}+\frac{\partial E_z}{\partial z} \qquad (13.26)$$

We separate out the symmetric and anti-symmetric parts of each matrix.

135

$$\sum_{\substack{\text{All algebras} \\ \text{Sym}}} [B,d]$$

$$= 4 \begin{bmatrix} 0 & 0 & 0 & 0 \\ 0 & 0 & 2\dfrac{\partial B_y}{\partial x}+2\dfrac{\partial B_x}{\partial y} & 2\dfrac{\partial B_z}{\partial x}+2\dfrac{\partial B_x}{\partial z} \\ 0 & 2\dfrac{\partial B_y}{\partial x}+2\dfrac{\partial B_x}{\partial y} & 0 & 2\dfrac{\partial B_z}{\partial y}+2\dfrac{\partial B_y}{\partial z} \\ 0 & 2\dfrac{\partial B_z}{\partial x}+2\dfrac{\partial B_x}{\partial z} & 2\dfrac{\partial B_z}{\partial y}+2\dfrac{\partial B_y}{\partial z} & 0 \end{bmatrix} \qquad (13.27)$$

$$\sum_{\substack{\text{All algebras} \\ \text{Anti}}} [B,d] =$$

$$4 \begin{bmatrix} 0 & 0 & 0 & 0 \\ 0 & 0 & \dfrac{\partial B_y}{\partial x}-\dfrac{\partial B_x}{\partial y} & \dfrac{\partial B_z}{\partial x}-\dfrac{\partial B_x}{\partial z} \\ 0 & -\dfrac{\partial B_y}{\partial x}+\dfrac{\partial B_x}{\partial y} & 0 & \dfrac{\partial B_z}{\partial y}-\dfrac{\partial B_y}{\partial z} \\ 0 & -\dfrac{\partial B_z}{\partial x}+\dfrac{\partial B_x}{\partial z} & -\dfrac{\partial B_z}{\partial y}+\dfrac{\partial B_y}{\partial z} & 0 \end{bmatrix} \qquad (13.28)$$

$$\sum_{\substack{\text{All algebras} \\ \text{Sym}}} \{E,d\} =$$

$$8 \begin{bmatrix} Div & \dfrac{\partial E_x}{\partial t}+\dfrac{\partial E_t}{\partial x} & \dfrac{\partial E_y}{\partial t}+\dfrac{\partial E_t}{\partial y} & \dfrac{\partial E_z}{\partial t}+\dfrac{\partial E_t}{\partial z} \\ \dfrac{\partial E_x}{\partial t}+\dfrac{\partial E_t}{\partial x} & Div & 0 & 0 \\ \dfrac{\partial E_y}{\partial t}+\dfrac{\partial E_t}{\partial y} & 0 & Div & 0 \\ \dfrac{\partial E_z}{\partial t}+\dfrac{\partial E_t}{\partial z} & 0 & 0 & Div \end{bmatrix} \qquad (13.29)$$

$$\sum_{\substack{\text{All algebras} \\ \text{Anti}}} \{E, d\} =$$

$$4 \begin{bmatrix} Div & \dfrac{\partial E_x}{\partial t} - \dfrac{\partial E_t}{\partial x} & \dfrac{\partial E_y}{\partial t} - \dfrac{\partial E_t}{\partial y} & \dfrac{\partial E_z}{\partial t} - \dfrac{\partial E_t}{\partial z} \\[3mm] -\dfrac{\partial E_x}{\partial t} + \dfrac{\partial E_t}{\partial x} & Div & 0 & 0 \\[3mm] -\dfrac{\partial E_y}{\partial t} + \dfrac{\partial E_t}{\partial y} & 0 & Div & 0 \\[3mm] -\dfrac{\partial E_z}{\partial t} + \dfrac{\partial E_t}{\partial z} & 0 & 0 & Div \end{bmatrix} \quad (13.30)$$

The electromagnetic inhomogeneous Maxwell equations:
Adding the two anti-symmetric parts of the tensor and forcing them into quaternion form leads to the inhomogeneous Maxwell equations of electromagnetism.

$$\frac{\partial E_x}{\partial t} - \frac{\partial E_t}{\partial x} = \frac{\partial B_z}{\partial y} - \frac{\partial B_y}{\partial z}$$

$$\frac{\partial E_y}{\partial t} - \frac{\partial E_t}{\partial y} = \frac{\partial B_x}{\partial z} - \frac{\partial B_z}{\partial x} \quad (13.31)$$

$$\frac{\partial E_z}{\partial t} - \frac{\partial E_t}{\partial z} = \frac{\partial B_y}{\partial x} - \frac{\partial B_x}{\partial y}$$

The GEM inhomogeneous Maxwell equations:
Summing the two symmetric matrices gives:

$$\sum_{\substack{\text{All algebras}}} \underset{\text{Sym}}{[B,d]} + \sum_{\substack{\text{All algebras}}} \underset{\text{Sym}}{\{E,d\}} =$$

$$8 \begin{bmatrix} Div & \dfrac{\partial E_x}{\partial t}+\dfrac{\partial E_t}{\partial x} & \dfrac{\partial E_y}{\partial t}+\dfrac{\partial E_t}{\partial y} & \dfrac{\partial E_z}{\partial t}+\dfrac{\partial E_t}{\partial z} \\[3mm] \dfrac{\partial E_x}{\partial t}+\dfrac{\partial E_t}{\partial x} & Div & \dfrac{\partial B_y}{\partial x}+\dfrac{\partial B_x}{\partial y} & \dfrac{\partial B_z}{\partial x}+\dfrac{\partial B_x}{\partial z} \\[3mm] \dfrac{\partial E_y}{\partial t}+\dfrac{\partial E_t}{\partial y} & \dfrac{\partial B_y}{\partial x}+\dfrac{\partial B_x}{\partial y} & Div & \dfrac{\partial B_z}{\partial y}+\dfrac{\partial B_y}{\partial z} \\[3mm] \dfrac{\partial E_z}{\partial t}+\dfrac{\partial E_t}{\partial z} & \dfrac{\partial B_z}{\partial x}+\dfrac{\partial B_x}{\partial z} & \dfrac{\partial B_z}{\partial y}+\dfrac{\partial B_y}{\partial z} & Div \end{bmatrix}$$

$$(13.32)$$

Forcing this into $\{SSA_{Anti}, SAS, ASS\}$ A_3 form gives:

$$\frac{\partial E_x}{\partial t}+\frac{\partial E_t}{\partial x} = \pm\left(\frac{\partial B_z}{\partial y}+\frac{\partial B_y}{\partial z}\right)$$

$$\frac{\partial E_y}{\partial t}+\frac{\partial E_t}{\partial y} = \pm\left(\frac{\partial B_x}{\partial z}+\frac{\partial B_z}{\partial x}\right) \qquad (13.33)$$

$$\frac{\partial E_z}{\partial t}+\frac{\partial E_t}{\partial z} = \pm\left(\frac{\partial B_y}{\partial x}+\frac{\partial B_x}{\partial y}\right)$$

We conclude:

$$\frac{\partial E_x}{\partial t}+\frac{\partial E_t}{\partial x}=0 \qquad \left(\frac{\partial B_z}{\partial y}+\frac{\partial B_y}{\partial z}\right)=0$$

$$\frac{\partial E_y}{\partial t}+\frac{\partial E_t}{\partial y}=0 \qquad \left(\frac{\partial B_x}{\partial z}+\frac{\partial B_z}{\partial x}\right)=0 \qquad (13.34)$$

$$\frac{\partial E_z}{\partial t}+\frac{\partial E_t}{\partial z}=0 \qquad \left(\frac{\partial B_y}{\partial x}+\frac{\partial B_x}{\partial y}\right)=0$$

Conclusion:

We have extracted a symmetric tensor which we assert is the energy-momentum tensor. This is half of the GR field equations.

As a bonus, we have the whole of classical electromagnetism as well.

Furthermore, we have three generations of matter.

We have seen that the homogeneous and inhomogeneous Maxwell equations of gravito-electromagnetism are all zero in the classical universe. If they were not zero, we would have a gravito-electromagnetic force in the classical universe. The cancellations that lead to these equations being zero depend on a balance between the $\{SSA^*, SAS, ASS\}$ A_3 algebras. We can speculate that, if this balance is disturbed by something, perhaps gravitational curvature of space-time, then an extra component of gravity would emerge from the imbalance in the form of gravito-electromagnetic forces. Could this be dark matter? We do not know.

Chapter 14

The Expanding Universe and Dark Energy

If we are to understand space-time, we need to understand why the empty space between galaxies is expanding. We have previously written of the expanding universe[53], but we briefly reproduce the essence of that exposition here before proffering an explanation of the observed acceleration of that expansion.

Consider the 2-dimensional space-time 'plane' which is the inside of a hyperbola (see below). The 'arms' of the hyperbola get further apart with increasing real variable. We associate the real axis with the age of the universe and the 'arms' of the hyperbola with the edges of the observable universe. As we progress along the real axis (we age) the edges of the universe get further apart and the space between those edges gets stretched.

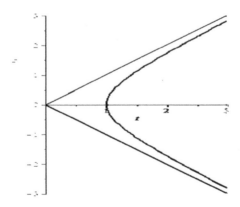

The units of the above graph are based upon the limiting velocity of the universe being unity, and the graph is squashed in the vertical direction.

Looking at the hyperbola, we have an inflationary start to the universe at time equals one unit. We see that the expansion approaches the exact limit, known as $\Omega = 1$.

---

[53] See: Dennis Morris The Physics of Empty Space

There is a problem with our explanation in that there is no sign of the accelerating expansion that cosmologists seem to have detected. We think this observed acceleration might be an optical illusion.

Consider an observer half a billion years after the start of the universe - 1st diagram below. As this observer looks at the universe, she looks back in time half a billion years and she looks through a distance of half a billion light years. On the diagram below, she looks back at 45° to the real axis, but she can see no further than half a billion light years. Since the universe is half a billion light years old, she thinks she is looking back to the start of the universe and so she thinks she is seeing the edge of the universe. Because the universe has expanded faster than the speed of light, she does not see all the way to the edge of the universe.

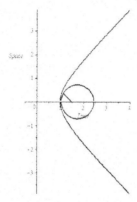

Now consider the observer's descendent looking at the universe 2 billion years after the start of the universe. Again, she cannot see all the way to the edge of the universe, but she believes she is seeing to the edge of the universe.

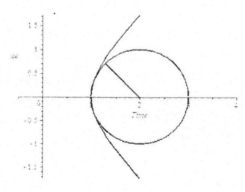

Now consider an observer who looks at the universe after three billion years. He can see three billion years into the past and so believes he is looking at the edge of the universe, but that edge is now nearer than 3 billion light years.

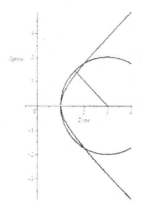

His descendent looking at the universe after 4 billion years will see the edge of the universe to be even further away than is actually the case.

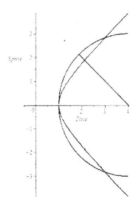

We see that the universe appears to be accelerating in its expansion for the later observers.

Your author does not know if the proffered explanation is the correct explanation, but it does avoid having to postulate some kind of universal dark energy driving the accelerating expansion. The reader will form their own opinion.

Chapter 15

Concluding Remarks

We have not provided a complete presentation of GR. We have made no mention of gravitational waves, of black holes or of the precession of planetary orbits. This subjects are very adequately presented elsewhere.

We believe we have explained from where our 4-dimensional space-time derives and why it has the nature it does have.

We have seen the curvature of our 4-dimensional space-time induced into our emergent expectation space by the local variation of the expectation A_3 phase. This is exactly the way the other forces of nature are derived in QFT.

We have seen the two spatially extensive classical forces, gravity and electromagnetism emerge together in one expectation field tensor. The other forces of nature are not spatially extensive; we conjecture that this is because they are not so closely connected to the emergent expectation space as is electromagnetism.

The whole of classical physics has emerged as 'expectation physics' from the spinor algebras and the spinor algebras derive from no more than the existence of the number one. Do we need an observer to do the 'expecting'? Your author thinks not. Classical physics seems to be one big average of spinor algebras.

There is a question concerning emergent expectation spaces which we have not previously addressed. How many of them are there?

We take the view that an emergent expectation space is not a geometric space in any meaningful way unless it has rotation within it. The only rotations we have are spinor rotations. The only spinor rotations which come through super-imposition are the two 2-dimensional rotations, $\mathbb{C}$ and $\mathbb{S}$, and two 3-dimensional rotations. These four types of rotation are each associated within their respective spinor algebras with a particular distance function. To have an emergent expectation space with rotation, we must have at least one of these distance functions as a sub-function of the expectation distance function that emerges from the super-imposition of isomorphic spinor algebras.

It is not known if there are any emergent expectation spaces other than our 4-dimensional space-time. Your author conjectures that our 4-dimensional space-time is the only emergent expectation space with rotation.

There is still much work to be done in this area of mathematics/physics. Each year, our knowledge in this area deepens, and your author apologises for being unable to present more than he has presented, but he hopes the reader has been given pause for thought from this book. Perhaps the reader will make advances in this area of theoretical physics. It is now time to return to the start of this book and re-read this book. Much will become clearer on a second reading.

Thank you for your time , it has been a pleasure writing for you.

Dennis Morris

Port Mulgrave

August 2015

Other Books by the Same Author

The Naked Spinor – a Rewrite of Clifford Algebra

Spinors exist in Clifford algebras. In this book, we explore the nature of spinors. This book is an excellent introduction to Clifford algebra.

Complex Numbers The Higher Dimensional Forms – Spinor Algebra

In this book, we explore the higher dimensional forms of complex numbers. These higher dimensional forms are connected very closely to spinors.

Upon General Relativity

In this book, we see how 4-dimensional space-time, gravity, and electromagnetism emerge from the spinor algebras. This is an excellent and easy-paced introduction to general relativity.

From Where Comes the Universe

This is a guide for the lay-person to the physics of empty space.

Empty Space is Amazing Stuff – The Special Theory of Relativity

This book deduces the theory of special relativity from the finite groups. It gives a unique insight into the nature of the 2-dimensional space-time of special relativity.

The Nuts and Bolts of Quantum Mechanics

This is a gentle introduction to quantum mechanics for undergraduates.

Quaternions

This book pulls together the often separate properties of the quaternions. Non-commutative differentiation is covered as is non-commutative rotation and non-commutative inner products along with the quaternion trigonometric functions.

The Uniqueness of our Space-time

This book reports the finding that the only two geometric spaces within the finite groups are the two spaces that together form our universe. This is a startling finding. The nature of geometric space is explained alongside the nature of division algebra space, spinor space. This book is a catalogue of the higher dimensional complex numbers up to dimension fifteen.

Lie Groups and Lie Algebras

This book presents Lie theory from a diametrically different perspective to the usual presentation. This makes the subject much more intuitively obvious and easier to learn. Included is perhaps the clearest and simplest presentation of the true nature of the Lie group $SU(2)$ ever presented.

The Physics of Empty Space

This book presents a comprehensive understanding of empty space. The presence of 2-dimensional rotations in our 4-dimensional space-time is explained. Also included is a very gentle introduction to non-commutative differentiation. Classical electromagetism is deduced from the quaternions.

The Electron

This book presents the quantum field theory view of the electron and the neutrino. This view is radically different from the classical view of the electron presented in most schools and colleges. This book gives a very clear exposition of the Dirac equation including the quaternion rewrite of the Dirac equation. This is an excellent introduction to particle physics for students prior to university, during university and after university courses in physics.

The Quaternion Dirac Equation

This small book (only 40 pages) presents the quaternion form of the Dirac equation. The neutrino mass problem is solved and we gain an explanation of why neutrinos are left-chiral. Much of the material in this book is drawn from 'The Electron'; this material is presented concisely and inexpensively for students already familiar with QFT.

An Essay on the Nature of Space-time

This small and inexpensive volume presents a view of the nature of empty space without the detailed mathematics. The expanding universe and dark energy is discussed.

Elementary Calculus from an Advanced Standpoint

This book rewrite the calculus of the complex numbers in a way that covers all division algebras and makes all continuous complex functions differentiable and integrable. Non-commutative differentiation is covered. Gauge covariant differentiation is covered as is the covariant derivative of general relativity.

Even Mathematicians and Physicists make Mistakes

This book points out what seems to be several important errors of modern physics and modern mathematics. Errors like the misunderstanding of rotation, the failure to teach the higher dimensional complex numbers in most universities, and the mathematical inconsistency of the Dirac equation and some casual errors are discussed. These errors are set in their historical circumstances and there is discussion about why they happened and the consequences of their happening. There is also an interesting chapter on the nature of mathematical proof within our society, and several famous proofs are discussed (without the details).

Finite Groups – A Simple Introduction

This book introduces the reader to finite group theory. Many introductory books on finite groups bury the reader in geometrical examples or in other types of groups and lose the central nature of a finite group. This book sticks firmly with the permutation nature of finite groups and elucidates that nature by the extensive use of permutation matrices. Permutation matrices simplify the subject considerably. This book is probably unique in its use of permutation matrices and therefore unique in its simplicity.

The Left-handed Spinor

This book covers the left-handed parts of mathematics which we call the chiral algebras. These algebras have CP invariance, violation of parity, and many other aspects which makes them relevant to theoretical physics. It is quite a revelation to discover that mathematics is left-handed.

Non-commutative Differentiation and the Commutator

(The Search for the Fermion Content of the Universe)

This book develops the theory of non-commutative differentiation from the fundamentals of algebra. We see what an algebraic operation (addition, multiplication) really is, and we discover that the commutator is a third fundamental algebraic operation within some division algebras. This leads to the first part of the derivation of the fermion content of the universe.

Bibliography

A First Course in General Relativity	Bernard F. Schutz
Clifford Algebras and Spinors	Pertti Lounesto
Cosmological Physics	John A. Peacock
General Relativity	Malcomb Ludvigsen
General Relativity	J. L. Martin
General Relativity	Robert M. Wald
General Theory of Relativity	P. A. M. Dirac
Gravitation (1973)	Misner C. W., Thorne K. S., Wheeler J. A.
Principles of Cosmology and Gravitation	M. V. Berry
Relativity	Wolfgang Rindler
Relativity Demystified	David McMahon
Quantum Field Theory	A. Zee
Quantum Theory Demystified	David McMahon
Space-time and Geometry	Sean M. Carroll

Index

1

T

U

W

Z

Made in the USA
Coppell, TX
16 July 2020

31140735R00092